Erica Thompson is a senior policy f
Economics' Data Science Institute
Mathematical Laboratory. With a P
has recently worked on the limita
spread, humanitarian crises, and climate change. She lives in West Wales.

Praise for *Escape from Model Land*

'A brilliant account of how models are so often abused and of how they should be used'

John Kay, author of *Other People's Money*

'A wise, lucid and compelling guide to how mathematical modelling shapes our world. Dr Thompson teaches us how to go from being unthinking consumers of models to sophisticated users, combining a rich variety of vivid examples and case studies with deep conceptual expertise'

Stian Westlake, CEO, Royal Statistical Society

'[A] healthy realism about data, algorithms and their limitations ... Thompson asks data scientists to be conscious of the choices and values in a model's design ... [offering] the basis for a constructive agenda'

The Economist

'Data, computing power, AI, and the models that use them will continue to proliferate. The wisdom, life experience, and humility to make the best use of those powerful tools will remain scarce. This delightfully wide-ranging book offers heaps of the latter to help us generate genuine insights from the former'

Charles J. Wheelan, bestselling author of *Naked Statistics*

'Demystifies the process of making the mathematical models that are increasingly used to make decisions about our lives . . . A thought-provoking and helpful guide for data scientists and decision makers alike'

Stephanie Hare, author of *Technology is Not Neutral*

'Offers a contemplative, densely encapsulated summary of her reflection and research . . . it's up to us to learn from models without being drawn in by their seductive elegance, and to ensure that the lessons from Model Land find substantive expression where it actually matters: in our messy, material, magnificent world'

Wall Street Journal

'Carefully researched and beautifully written . . . For an open-minded reader keen to expose, understand and potentially reconstruct their own worldview, *Escape from Model Land* is, at the same time, an uncomfortable and uplifting read. It shines a gentle light on many of our own norms and beliefs'

Kevin Anderson

'An eye-opening account . . . Thompson offers a host of lessons . . . The result is a thoughtful, convincing look at how data works'

Publisher's Weekly

'Brilliant . . . a highly engaging work of popular science'

E&T Magazine

Escape from Model Land

How Mathematical Models Can Lead Us Astray and What We Can Do About It

ERICA THOMPSON

LONDON

First published in Great Britain in 2022 by Basic Books UK
An imprint of John Murray Press

This paperback edition published in 2023

1

Figure 2: Photograph of Bill Phillips from LSE Library

A CIP catalogue record for this title is available from the British Library

Paperback ISBN 9781529364897
ebook ISBN 9781529364903

Typeset in Janson Text LT by Hewer Text UK Ltd, Edinburgh
Printed and bound in Great Britain by Clays Ltd, Elcograf S.p.A.

John Murray Press policy is to use papers that are natural, renewable
and recyclable products and made from wood grown in sustainable
forests. The logging and manufacturing processes are expected to
conform to the environmental regulations of the country of origin.

Carmelite House
50 Victoria Embankment
London EC4Y 0DZ

www.basicbooks.uk

John Murray Press, part of Hodder & Stoughton Limited
An Hachette UK company

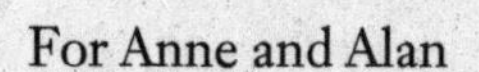

For Anne and Alan

Contents

I

Locating Model Land

> I would like to know if anyone has ever seen a natural work of art. Nature and art, being two different things, cannot be the same thing. Through art we express our conception of what nature is not.
>
> Pablo Picasso, *Picasso Speaks* (1923)

Our world is awash with data. Satellites continuously watch the Earth below, sending back data about rainforest destruction in the Amazon, cyclones in the Indian Ocean and the extent of the polar ice sheets. Tracking algorithms continuously watch human behaviour online, monitoring our time spent on different websites, measuring our propensity to click on different adverts and noting our purchases with an eagle eye. Athletes from Olympians to casual joggers track their daily performance and physical condition. Financial markets create vast quantities of high-frequency price data.

These are all observed directly from the real world, but without a framework to interpret that data it would be only a meaningless stream of numbers. The frameworks we use to interpret data take many different forms, and I am going to refer to all of them as models. Models can be purely conceptual: an informal model of the casual jogger might be 'when my heart rate goes up I am exercising more intensely'. But with data, we can go further and make quantitative models that seek to describe the relationships between one variable and another: '10% of people who clicked on this product listing for a bicycle pump in the last week went on to buy it.' The contribution of the model is to add *relationships* between data. These are observed relationships, and they may have some uncertainty. We do not yet know *which* 10% of customers would buy the pump. They are also influenced by our own expectations. We did not test how many of the people who clicked on the bicycle pump went on to buy an ironing board or a tin of tomatoes, but we might want to test how many of the people who clicked on this bicycle pump went on to buy a competing brand of pump, or whether those people also bought other sports equipment.

And what is the point of doing that? Well, very few people are interested in these kinds of numbers for their own sake. What the vendor *really* wants to know is whether the purchase of the bicycle pump can be made more likely, and whether other items could be advertised as 'related' in order to interest the customer in spending more money on the same site. What the funding agencies who send up satellites want to know is how we can understand and predict weather and climate more effectively in order to avoid negative impacts. What the athlete wants to know is how they can improve their performance. What the financial trader wants to know is how they can get an edge.

As the first wave of the Covid-19 pandemic swept across the world in early 2020, governments desperately sought advice from epidemiologists and mathematical modellers about what might happen next and how to respond effectively. Information was scarce, but decisions were urgent. There was uncertainty about the characteristics of the novel virus, uncertainty about how people and communities would respond to major government interventions, and huge uncertainty about the longer-term epidemiological and economic outcomes.

Into that context, mathematical modellers brought quantitative scenarios – possible futures that might come to pass depending on certain assumptions. In particular, these models showed that if the proportion of serious cases were in the plausible range and no interventions were made, then a large number of people could be expected to require hospitalisation simultaneously. The very simplest model of infection transmission from one person to more than one other person is a framework with which you can make both large-scale conclusions about exponential growth in a population and small-scale conclusions about reduction of personal risk by reducing contacts. More complex models can represent semi-realistic differences in contact patterns over time for different sectors of society, and account for the effects of new information, better treatments, or vaccination programmes. Even the sad human toll can only be estimated through modelling. Doing so offers the possibility of predicting the effect of different kinds of interventions, obviously of interest to large-scale policy-making. And so in the years since their introduction, epidemiological models have held a place at the top table of national policy-making and on the front pages of the public media.

In short, the purpose of modelling relationships between data is to try to predict how we can take more effective actions in future in support of some overall goal. These are real-world questions informed by real-world data; and when they have answers at all, they have real-world answers. To find those answers, we have to go to Model Land.

Entering Model Land

There are no border controls between the real world and Model Land. You are free to enter any time you like, and you can do so from the comfort of your armchair simply by creating a model. It could be an ethereal, abstract conceptual model, a down-to-earth spreadsheet or a lumbering giant of a complex numerical model. Within Model Land, the assumptions that you make in your model are literally true. The model *is* the Model Land reality.

Of course, there are many different regions of Model Land, corresponding to the many different models we can make, even of the same situations. In one part of Model Land, the Earth is flat: a model to calculate the trajectory of a cricket ball does not need to include the radius of the Earth. In another part of Model Land, the Earth is a perfect sphere: a globe shows the relative positions of countries and continents with sufficient accuracy for most geographical purposes. To other modellers, the Earth is an oblate spheroid,

43km wider at the equator than through the poles. For the most detailed applications like constructing a GPS system or measuring the gravitational field, we also need to know about mountains and valleys, high plateaus and ocean trenches. All of these are models of the same Earth, but with different applications in mind.

Model Land is a wonderful place. In Model Land, because all of our assumptions are true, we can really make progress on understanding our models and how they work. We can make predictions. We can investigate many different configurations of a model, and run it with different inputs to see what would happen in different circumstances. We can check whether varying some characteristics of the model would make a large or a small difference to the outcomes. The assumptions that underlie our statistical methods are also true in Model Land, and so our statistical analyses can be unencumbered by caveats.

That's not to say that life is easy in Model Land. We still have to make the model and 'do the math'. Whole careers can be spent in Model Land doing difficult and exciting things.

Model Land is not necessarily a deterministic place where everything is known. In Model Land, there can still be uncertainty about model outcomes. The uncertainty might arise from our limited ability to measure quantities or initial conditions, or from difficulties in defining model parameters, from randomly determined elements in the model itself, or from the chaotic divergence of model trajectories, also known as the Butterfly Effect.

In Model Land, these uncertainties are quantifiable. Or at least they are quantifiable in principle, if one had a sufficiently large computer. With a smaller computer we can estimate the uncertainties and, given further information, the estimates we make will converge on the right answer.

In the gap between Model Land and the real world lie unquantifiable uncertainties: is this model structurally correct? Have I taken account of all relevant variables? Am I influencing the system by measuring and predicting it?

Most fields distinguish between these two kinds of uncertainty. In economics, the quantifiable uncertainties are often referred to as

Risk and the unquantifiable as Uncertainty; in physics, confusingly, the exact opposite terminology is sometimes used and uncertainty generally refers to something that can be quantified. Unquantifiable uncertainties are variously known as deep uncertainty, radical uncertainty and epistemic uncertainty.

Let's stay in Model Land for now. Think of a billiards table and the uncertainty involved in predicting where a ball will end up when you hit it. In principle, this is a simple calculation involving energy, friction and deceleration along a straight line. In practice, though, even in Model Land we can recognise the limitations of a single calculation. There is a measurement uncertainty because you cannot define exactly where the ball is to start with. There is an initial condition uncertainty because you cannot define exactly how much energy you will impart to the ball when you hit it. There is a boundary condition uncertainty because you cannot define exactly how much energy the ball will lose when bouncing off the edges. And these, taken together, result in a random uncertainty about the exact position of the ball when it comes to rest. In principle, you could define a range of plausible conditions for each of these quantifiable unknowns and use them to calculate a range of plausible resting points for the ball after the shot. With luck, when the experiment is tried, the ball would be somewhere near the middle of that range. Physicists would refer to the plausible range as the 'error bars' of the calculation; statisticians might call it a 'confidence interval'.

Deep or radical uncertainty enters the scene in the form of the unquantifiable unknowns: things we left out of the calculation that we simply could not have anticipated. Maybe it turns out that the billiards table was actually not a level surface, your opponent was blowing the ball off course or the ball fell into a pocket rather than bouncing off the edge. In that case, your carefully defined statistical range of projected outcomes could turn out to be completely inadequate. These are also called epistemic uncertainties, from the Greek *episteme* ('knowledge'), because they reflect the limits of our knowledge and the limits of our ability to predict. And yet they are not completely inaccessible: we can list these uncertainties, study them, quantitatively or qualitatively evaluate the likelihood of their occurrence, even take real-world steps to prevent them. Yes, there could be truly unexpected events – Black Swans – by which we might be blindsided. But the majority of surprising events are not really Black Swans. Some people saw the 2008 financial crisis coming. Pandemics have been at the top of national risk registers for decades. Climate tipping points are absolutely on the radar of mainstream scientific research. It's not that we think these kinds of events can't happen, it's that we haven't developed an effective way of dealing with or formalising our understanding that they *could* happen. One premise of this book is that unquantifiable uncertainties are important, are ubiquitous, are potentially accessible to us and should figure in our decision-making – but that to make use of them we must understand the limitations of our models, acknowledge their political context, escape from Model Land and construct predictive statements that are about the real world.

The problems of risk and uncertainty are not the only things that bedevil life in Model Land. Although scientists prefer to think of models as being confined to a scientific arena of facts, data and possibility, the degree to which models are unavoidably entangled with ethics, politics and social values has become very clear. Narratives about risk and responsibility are attached to them. Ethical judgements are inherent in the things that are represented and in the things that are not. Models are tools of social persuasion and vehicles for political debate.

We see this in the models for the Covid-19 pandemic when we considered who should stay home, who should wear masks and who should have a vaccine, or how we should answer a hundred other questions. We also see this in an area I am more familiar with from my own previous research: climate change. Again, the science itself is only a small part of the story. As our climate continues to move gradually but measurably away from the weather patterns of the twentieth century, the greenhouse effect – an unfortunate fact of life, like person-to-person viral transmission – is no longer in debate. But the incredibly complex political question of what exactly should be done about it remains, and is again mediated by complex mathematical models. Both models of the physical climate and models of the economic system inform opinions about what kind of target to aim for and the trade-offs between the costs of climate action and the costs of climate inaction. As with Covid, politics enters into the models through value judgements about what kinds of losses and damages are acceptable, the distribution of risk across different geographical communities and the question of who should take on the costs of doing something about it. For either climate or Covid, the precautionary principle is not much help: we can be precautionary about the emerging risk, or precautionary about taking upfront economic hits before the harm is truly known. There are many analogies between responses to fast-emerging Covid risks and slow-emerging climate risks from which one could draw many conclusions. But a priority has to be getting a better grip on the way that mathematical and scientific models interact with society and inform our decision-making structures.

You cannot avoid Model Land by working 'only with data'. Data, that is, measured quantities, do not speak for themselves: they are given meaning only through the context and framing provided by models. Nor can you avoid Model Land by working with purely conceptual models, theorising about dice rolls or string theories without reference to real data. Good data and good conceptual understanding can, though, help us to escape from Model Land and make our results relevant once more in the real world.

Escaping from Model Land

Though Model Land is easy to enter, it is not so easy to leave. Having constructed a beautiful, internally consistent model and a set of analysis methods that describe the model in detail, it can be emotionally difficult to acknowledge that the initial assumptions on which the whole thing is built are not literally true. This is why so many reports and academic papers about models either make and forget their assumptions, or test them only in a perfunctory way. Placing a chart of model output next to a picture of real observations and stating that they look very similar in all important respects is a statement of faith, not an evaluation of model performance, and any inferences based on that model are still in Model Land.

One aim of this book is to encourage modellers – and we are all modellers – to think more critically about model evaluation: how is it, really, that we can know that our models are good enough predictive tools to tell us what might happen in the real world? The defeatist answer to this question is that we can never know. Although that might be technically correct, there are plentiful examples of wildly successful mathematical modelling in the history of science and engineering upon which we justifiably depend and by which we have not been failed. In my view, the continued success of modelling depends on creating a programme of understanding that uses models as a tool and as a guide for thinking and communication, and that recognises and is clear about its own limits.

As such, one priority is to understand the exits from Model Land and signpost them more clearly; this is something I will do in subsequent chapters. Briefly, there are two exits from Model Land: one quantitative and one qualitative. The quantitative exit is by comparison of the model against out-of-sample data – data that were not used in the construction of the model. This is more difficult to arrange than you might imagine: out-of-sample data can be hard to come by or take a long time to be collected. But short-timescale predictive engines like weather forecasts can take the quantitative exit very effectively, repeatedly forecasting and comparing forecasts with observations to build up a statistical picture of when and where the model is accurate, and when and where it struggles. This is best-practice evaluation, but simply not possible for many important classes of models, such as financial models and longer-term climate models. For financial models, although we may have a stream of out-of-sample data, the problem is that the underlying conditions may be changing. The standard disclaimer on investment opportunities also applies to models: past performance is no guarantee of future success.

The qualitative exit from Model Land is much more commonly attempted, but it is also much more difficult to make a successful exit this way. It consists of a simple assertion, based on expert judgement about the quality of the representation, that the model bears a certain relationship with the real world. An example of failure to escape using this route is the above characterisation of reports that put a picture of model output next to a picture of the real world and assert that they look sufficiently similar that the model must therefore be useful for the purpose of whatever prediction or action the reporters are discussing. This is essentially an implicit expert judgement that the model is perfect; Model Land is reality; our assumptions are either literally true or close enough that any differences are negligible. As we will see, this kind of naive Model Land realism can have catastrophic effects because it invariably results in an underestimation of uncertainties and exposure to greater-than-expected risk. The financial crisis of 2008 resulted in part from this kind of failure to escape from Model Land. A more successful escape attempt is embodied in the language used by the Intergovernmental Panel on Climate Change (IPCC). When discussing results from physical climate models, they systematically reduce the confidence levels implied by their models, based on the best expert knowledge of both the models and the physical processes and observations those models claim to represent. This results in a clear distinction between Model Land and the real world. Better still, the statement attaches an approximate numerical magnitude to this distinction, which is relevant for real-world decision-makers using this information and for other modellers.

The acknowledgement of such uncertainty should give us, as information consumers, greater confidence in the ability of these authors to incorporate some judgements from outside Model Land into their projections, and therefore greater confidence that those projections will prove trustworthy.

The subjectivity of that second escape route may still worry you. It should. There are many ways in which our judgement, however expert, could turn out to be inadequate or simply wrong. As Donald Rumsfeld, former US secretary of defense, put it: there are 'known unknowns' and there are 'unknown unknowns'. If the known knowns are things that we can model, and the known unknowns are things that we can subjectively anticipate, there will always be unknown unknowns lurking in the darkness beyond the sphere of our knowledge and experience.

But how else can we account for them? By definition, we cannot infer them from the data we have, and yet we must still make decisions. It would not make sense for the unknown magnitude of unknowns to be the main driver of all our decisions: in that case, we would all be living in constant fear of a nuclear war, economic collapse or (another) viral pandemic. On the other hand, discounting these possibilities completely is a recipe for another kind of disaster.

Models and machines are not good at living with these kinds of unquantifiable uncertainties. Either they require that we quantify the unquantifiable or they ignore it. When there are infinities or undefinables, they cannot work at all. Yet our human brains are supremely good at this task. Without calculation, we can integrate many opposing sources of information and multiple conflicting goals, and come out with not just a single decision, but a narrative about why that decision has been reached and what it aims to achieve; a narrative that can then be put to use in communicating with others. Models struggle to emulate this kind of decision-making fluency.

But my argument is not that we should throw away the models. Models are an inseparable and vital part of modern decision-making systems. What I want to argue here is that the human brain is also an inseparable and vital part of modern decision-making systems. Going further, this is a partnership: the human brain is responsible for constructing models; models provide quantitative and qualitative insights; the brain can integrate these with other, non-modelled insights; and the upshot is a system that can be better than either brain or model acting alone.

Or it can be worse. Both model thinking and human thinking are notably subject to standpoint biases. If I am constructing models from a Western, Educated, Industrial, Rich, Developed ('WEIRD') standpoint, then both my brain and my model will contain WEIRD assumptions that I may not even notice. As such, it is a key argument of this book that diversity in models, and (real) diversity in modellers, can provide greater insight, improved decision-making capacities and better outcomes. Mathematical modelling in particular is a very WEIRD pursuit, so it deserves greater scrutiny for the overall effectiveness and the outcomes that are served by prioritising modelling approaches over alternative forms of decision support.

As I will show, reliance on models for information tends to lead to a kind of accountability gap. Who is responsible if a model makes harmful predictions? The notion of 'following the science' becomes a screen behind which both decision-makers and scientists can hide, saying 'the science says we must do X' in some situations and 'it's only a model' in others. The public are right to be suspicious of the political and social motives behind this kind of dissimulation. Scientists and other authorities must do better at developing and being worthy of trust by making the role of expert judgement much clearer, being transparent about their own backgrounds and interests, and encouraging wider representation of different backgrounds and interests.

Using models responsibly

I started thinking about these questions as a PhD student in Physics ten years ago, when I conducted a literature review on mathematical models of North Atlantic storms. I found that there were many models, predicting different overall effects and producing contradictory results, all peer-reviewed and published, and with conflicting rather than overlapping uncertainty ranges. I realised that I hadn't learned very much about North Atlantic storms, but I had learned a lot about how we make inferences from models, and I was worried about what that meant for confidence in other modelling studies. Since then, I have tried to find a balanced way to proceed, in between what I think are two unacceptable alternatives. Taking models literally and failing to account for the gap between Model Land and the real world is a recipe for underestimating risk and suffering the consequences of hubris. Yet throwing models away completely would lose us a lot of clearly valuable information. So how do we navigate the space between these extremes more effectively?

I have a feeling there are a lot of people asking similar fundamental questions. Not just in science, but those who are doing any kind of quantitative analytics: actuaries, financial traders, energy forecasters, start-up founders, humanitarian responders, public health analysts, environmental consultants, social-media marketers, bookies, sports teams – you name it! Models are ubiquitous now in everyday life, analysing and interpreting rivers of data gathered in many different ways from personal devices to satellites. If 'data is the new oil', then models are the new pipelines – and they are also refineries. How are models constructed? To whom do they deliver power? How should we regulate them? How can we use them responsibly?

I do not have final answers to these questions, but I hope that this book can help to address them: first, by reframing generic worries about models into more specific and tractable questions; second, by emphasising that the social and political content of models is at least as important as their mathematical and statistical content; and third, by offering a guide to some of the routes by which we can escape from Model Land.

2

Thinking Inside the Box

> There are more things in heaven and earth, Horatio,
> Than are dreamt of in your philosophy.
>
> William Shakespeare, *Hamlet* (1602)

If we are going to fit our complex world into a box in order to be able to model it and think about it, we must first simplify it. This is impossible to avoid, since we have no replica Earth on which to experiment. Inside the box, a certain set of rules apply and certain types of analysis can be performed. The box itself may have particular characteristics, and be able to hold some items and not others. There are many different ways to simplify, and hence many different models. But because of the simplification, each model can only be used to ask certain types of questions, and each model is only capable of offering certain kinds of answers.

Consider a spherical cow

A physicist is called in by a dairy farmer to see if the milk yields on the farm can be improved with the application of science. The physicist takes a look around the farm, asks lots of questions, makes notes and promises the farmer that indeed much progress can be made. Two weeks later the farmer eagerly opens a long report and begins to read: 'Consider a spherical cow of radius 1, in a vacuum . . .'

This old joke trades on the well-known tendency of scientists to simplify situations in order to make them tractable. Although the example here is certainly apocryphal, there are probably many real-life situations where a scientific consultant has been called in for some such task, done immense calculations on a simplified version of the question in hand and presented a detailed answer that is perfectly correct in Model Land – but totally useless to the client because it ignores some practical reality.

On the flip side, of course, we can imagine another report that the physicist might have presented. Perhaps they decided to model not a spherical cow but a highly realistic cow, taking many years and multiple supercomputers to simulate the workings of the digestive system, with teams of researchers in different disciplines looking at hoof and hair growth, lactation and food consumption. The model would be calibrated using a detailed observational study of lab animal Daisy and forward-project the impacts of changing Daisy's dietary routine. Results from this research programme, accompanied by multiple caveats about uncertainty and its inapplicability to other breeds of cow or non-laboratory situations, may again be of very limited use to the bemused (and probably now-retired) farmer.

The advantage of a mathematical model is the ability to neglect any aspects of the situation that are not immediately relevant, tractable or important. By removing these aspects, we can reduce the problem to its essence, highlighting only the causal links and developing insight into the behaviour of the subject. That is a best-case scenario. The disadvantage of a mathematical model is the necessity of neglecting any aspects of the situation that are not immediately relevant, tractable or important. By removing these aspects, we remove the problem from its context, potentially losing the information that could help us to understand the behaviour or even making the model totally wrong. That is a worst-case scenario.

The art of model-making is to draw the boundaries sufficiently wide that we include the important contributing factors, and at the same time sufficiently narrow that the resulting model can be implemented in a useful way. The exact location of each of those boundaries varies, and in some cases it may be that they do not even overlap.

The important factors will change depending on what kind of question you are asking and on how much tolerance you have for uncertainty in the answer.

The parameters of a model are essentially the control knobs that can be turned one way or the other to give a different output. If we model some data points as a straight line, there are two parameters: the slope of the line and its starting point. In order to fit the model to the observed data, we adjust the slope and the starting point until the line is as close as possible to the data points. This is what is termed linear regression, and there are simple statistical methods for automating the process. More complicated models may have many thousands of parameters with much more complicated effects, but the process of fitting or calibrating the model to the observed data remains the same: change the parameters until you get a model output that is closest (in some way) to the observations.

The more parameters a model has, in general, the more control we have over its behaviour and the more opportunity the model has to fit the data. Hungarian-American polymath John von Neumann is reputed to have said, 'With four parameters I can fit an elephant. With five I can make him wiggle his trunk.' The implication is that if we can fit anything, then the model has no explanatory power. If we can fit nothing, of course, it equally has no explanatory power. We gain confidence in a model by being able to fit the observations without going through great contortions to do so, because this shows that the variation in the observations is in some sense contained within the simple principles expressed by the model. This principle of simplicity is also termed 'Ockham's Razor', after a thirteenth-century English monk, William of Ockham, who wrote extensively on theological and philosophical matters. One of his arguments was essentially that God is the simplest explanation possible for the existence of things, and therefore it is more reasonable to believe in divinely revealed truths than in a multiplicity of other causes.

Ockham's Razor has inspired many other maxims and methodologies, from 'Keep It Simple, Stupid' to advanced statistical concepts like the Akaike Information Criterion. As computing power has become larger and larger over the last century, there has been greater opportunity to construct large and complicated models with lots of parameters. Because of this proliferation of parameters, there is a great need for careful thought about how to measure the quality of models and the informativeness of the 'answers' that they help to generate. In practice, the complexity of a model is usually determined less by abstract principles and more by the limit of the modelling materials you have to hand, be that a piece of paper, a supercomputer or a sack of assorted plumbing.

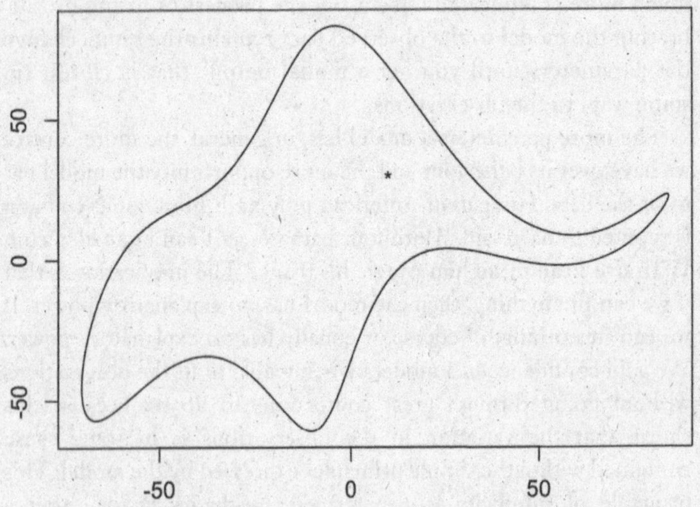

Figure 1: In fact, although several people have tried to fit an elephant with four parameters, this turns out to be somewhat difficult. The closest attempt uses four complex parameters (effectively eight parameters) to fit something that is vaguely recognisable as an elephant, and a fifth complex parameter (effectively ten parameters) to 'wiggle the trunk'. Reproduced using TinyElephant.R code kindly provided by Hans van Beek (2018), based on the equations of Mayer, Khairy and Howard (2010).

A hydraulic economy

Bill Phillips scraped a pass in his Sociology and Economics degree from the London School of Economics (LSE) in 1949, having spent much of his free time in the previous year engineering a physical model of the British economy using water moved between tanks with a system of pipes, pumps, floats and levers. The Newlyn-Phillips machine, also called the MONIAC (Monetary National Income Analogue Calculator), conceptualises and physically represents money as liquid, created by credit (or export sales) and circulated around the economy at rates depending on key model parameters like the household savings rate and levels of taxation.

The key idea here is that there are 'stocks' of money in certain nominal locations (banks, government, etc), modelled as tanks that can be filled or emptied, and 'flows' of money into and out of those tanks at varying rates. The rate of flow along pipes is controlled by valves which in turn are controlled by the level of water in the tanks. Pumps labelled 'taxation' return water into the tank representing the national treasury. The aim is to set the pumps and valves in a way that allows for a steady-state solution, i.e., a closed loop that prevents all of the water ending up in one place and other tanks running dry. In this way a complex system of equations is represented and solved in physical form.

Economic historian Mary Morgan describes how the model helped to resolve a debate when the model was first demonstrated to an academic audience in November 1949:

> When the red water flowed around the Newlyn-Phillips machine on that day, it resolved for the audience a strongly fought controversy in macroeconomics. In simplified form, Keynesians argued that the interest rate is determined by liquidity preference: people's preferences for holding stocks of money versus bonds. Robertson argued that the interest rate is determined by the supply and demand for loanable funds: primarily the flow of savings versus that of investment. When stocks and flows really work together – as they did that day in the machine

> demonstration – it became clear that the theories of Robertson and Keynes were neither inconsistent nor alternative theories but rather were complementary, but more important – they had been integrated in the machine's economic world.

Interestingly, this machine does not seem to have been much used directly for predictive purposes, but performed a more pedagogical role in illustrating and making visible the relationship between different aspects of the economy. The physical, visible representation of the abstract equations used by macroeconomists gave a structure for thinking about those equations in a new way, making new links and drawing new conclusions about the relationships. It showed, for example, the time lag required for a new balance to be achieved upon making some intervention (such as a change to taxation), since the water levels needed to propagate all the way around the system to find a new steady state. By contrast, the equations solved on paper directly for the equilibrium state could not show these short-term effects. And yet if Phillips had had access to a supercomputer, it is unlikely that he would have come up with this innovatively different perspective at all.

The scope of any model is constrained by the nature of the model itself. The Phillips-Newlyn machine provided a new and valuable perspective, but it also inherited biases, assumptions and blind spots from the pre-existing equation set it was designed to represent. Only certain kinds of questions can be asked of this model, and only certain types of answers can be given. You can vary the parameter called 'household savings rate' and observe the 'healthcare expenditure', but you cannot vary a parameter called 'tax evasion' or observe 'trust in government'. The kinds of question that are inside and outside the scope of a model are part of the nature of a model, determined by the priorities and interests of whoever made the model initially and probably also partly determined by mathematical (or, in this case, hydraulic) tractability. It may seem obvious, but there is no point in trying to ask a model about something that is outside its scope. Just as you cannot ask the Newlyn-Phillips machine about tax evasion, you also cannot ask a physical climate model about the most appropriate price of carbon, nor can you ask an epidemiological model whether young people would benefit more from reduced transmission of a virus or from access to in-person education. But when you do ask about something within scope, you need to bear in mind that the answer can only come in certain forms. The only way that the Newlyn-Phillips machine can represent economic failure, for example, is by the running-dry of the taxation pumps; there is no concept of political failure by imposing too-high taxation or failing to provide adequate public services. And the only success is a continued flow of money.

With the tools at hand limiting the kinds of representations that we make, it is important to understand that there is no rigid hierarchy by which a supercomputer must necessarily be a 'better' tool than a sack of plumbing. Each choice of representation lends itself to certain kinds of imaginative extension – which we will look at in the next chapter – and also inevitably limits the range of things we can use it for.

Figure 2: Bill Phillips in the 1960s, demonstrating the use of his hydraulic model of the economy.

The ad hoc standard

Models become a standard for ad hoc reasons: perhaps by being the first to be used or developed or to be made publicly available, or as a result of having led to particularly good or quick results. In climate science, which typically uses very complex models run on supercomputers, the HadCM3 model developed by the UK's Hadley Centre for Climate Research has been a workhorse of the global climate community for some years. It has been described by eminent climate scientist Isaac Held as the 'E. coli of climate models'. (This was definitely intended as a compliment, as when Garry Kasparov referred to chess as a 'drosophila [fruit fly] of reasoning'.) It was one of the first climate models not to require a so-called 'flux correction' or boundary adjustment to fit observed data, and one of the first to be made available more widely to download and run. As computing power has advanced, it is now able to be run on small personal computers rather than the supercomputers that today's state-of-the-art models require.

Models of this type provide opportunities for experimentation with climate modelling by researchers who do not have access to state-of-the-art models. Given that the latter almost by definition push up against the boundaries of currently available computing power, it is usually impossible to do full model runs (around a hundred simulated years) even thousands of times within a couple of (real) years of their development. By contrast, HadCM3 must by now have been run at least millions of times. This collective experience with the model gives us both the numbers on which to perform robust statistical analyses of model behaviour, and the variation of studies through which to get a picture of its strengths, weaknesses and quirks. One of the greatest benefits is that the shorter run times of simpler models like HadCM3 allow for much greater exploration of the sensitivity of the model to altered inputs (or parameters, or assumptions) than is possible with the latest models.

Nevertheless, projects typically only use HadCM3 because more sophisticated models are unavailable or impractical. And research funding to continue to develop the understanding of HadCM3 for its own sake would be a very difficult ask. In biology, researchers are interested in other organisms for their own sake and not solely as stand-ins, of course, whereas in climate science we are only interested in the models to the extent that they can tell us something meaningful about the real Earth system. Because it is widely believed that models with more detailed physical representations will be more realistic in their outputs (and because the continued increase in computing power allows it), there is a general progression towards more detailed models superseding all but a few of the older, less detailed versions.

Still, the continual process of supersedure leaves us with some difficult questions about superseded models: if they ever told us something, what was it? If they had great value when they could only be done on a supercomputer, why do they have lesser value and interest now that they can be run many more times, much more easily? What will happen to the current model when it is necessarily superseded by the next? If the scope of the model determines the kinds of questions that can be asked and answered, does further development result in a more useful model?

If the model is not being used to inform a decision, we can avoid talking about these questions. We can simply say that the most advanced models are able to help us explore our most complex musings, regardless of their absolute quality. Or we can hope that the continual evolution of models is moving towards some perfect-model endpoint, a true Theory of Everything that will be able to answer all our questions about the future with some known degree of accuracy. This hope will be in vain, though: it's not generally the case that the most complex models are the best, except where they are used naively to 'fit an elephant', and we have already seen that just fitting the past-elephant will not necessarily help us to understand more about a future-elephant.

Adequacy-for-purpose

The key distinction here is not whether a model is the 'best available' at the present time, but, as philosopher Wendy Parker has emphasised, whether it is 'adequate for purpose'. If we define what we mean by 'best', there will be exactly one best model, but there could be many that are adequate for a given purpose, or just one, or perhaps none at all.

For example, if I want to make a decision about whether to take an umbrella when I go out later, then a look at any weather forecast is an adequate decision-making tool – by which I mean that I can make a confident decision based on that information. It doesn't mean absolute certainty: the forecast might, for instance, say that there is a 30% chance of rain. In that case, if I were going to a garden party I would probably take an umbrella, but if I were just popping to a local shop I might choose to risk the possibility of wet clothes for the benefit of not having to carry the umbrella. There are multiple forecast-providers that could give me this information. They generally give me very similar forecasts, and I would be able to make an informed decision that reflects my own risk attitude. Occasionally, unless I choose the precautionary approach of always taking an umbrella regardless of forecast, I will get caught in the rain.

By contrast, if I wanted to decide now whether to take an umbrella to 'Auntie's garden party three Sundays from now', the detailed weather forecast would be no use at all: it is not adequate for this purpose. But I am not left completely in the dark. First, I might fall back on my expectations of the typical weather in June, or in December, which are generated by personal experience rather than by the forecast model. Second, I might prioritise my risk attitude as a decision input and choose to take an umbrella to the garden party as a precaution even though I expect the weather in June to be fine.

Where we are making important decisions that affect the lives of many people, it makes sense to have some idea of what we mean by a model that is adequate for the purpose of informing those decisions. Understanding and quantifying that adequacy-for-purpose is what I mean by escaping from Model Land, and the rest of this book looks at the mathematical, philosophical and social challenges of doing so. If the same model has been used successfully to inform constructive decision-making many times in the past, then that is a good indication (though not a guarantee) that it may be successful in future. Conversely, if it has been used in the past unsuccessfully, we can evaluate its lack of adequacy-for-purpose and steer clear of it – also a useful outcome. The difficult question that Parker and other philosophers have struggled with is what to do when you don't have a comprehensive and directly relevant archive of either past success or past failure, such as for climate models, economic models or epidemiological models. Then, in addition to the limited available data, adequacy-for-purpose has to be evaluated with respect to indirect criteria such as whether the model describes the relevant real-world phenomena sufficiently, whether it represents well-confirmed theory (such as the laws of physics) reasonably well or whether it otherwise 'looks plausible' to a panel of experts. This has benefits and challenges, both of which we will explore.

Of mice and models

Biological model organisms are used for their comparative simplicity, tractability and ethical acceptability. When one animal is used as a model for another, some questions become in-scope and others move out-of-scope. Genetic studies of inheritance prefer an organism with relatively short generation times so that many experiments can be performed in the time it takes to do a PhD: so we might choose fruit flies. We cannot perform a controlled trial of the effect on humans of new and relatively unknown substances, but it is (currently) somewhat more acceptable to do so on mice.

The use of model organisms in biology has a long history, from *Arabidopsis* (a mustard plant) to zebrafish by way of the fruit fly, the E. coli bacterium and the classic laboratory mouse. One advantage of the model-organism concept is that it provides a way to standardise studies across different laboratory settings. A lab in California and a lab in Beijing can conduct essentially the same experiment on the 'same' model mouse and – hopefully – replicate their results. Working with a common model, the global research effort can build upon each result to paint a fuller picture of the workings of that particular species, with some confidence that the statements made by one lab are directly comparable with statements made by another.

It is also quite clear that mice are not humans. Most people understand that effects on mice, while useful and suggestive of plausible results in humans, do not offer any guarantee of the same – although, to read the news, you might sometimes think otherwise. Dramatic media headlines accompany research results showing progress on high-profile health challenges like cancer, diabetes, AIDS, heart disease and so on. Often the headline does not include the key information that the study was performed 'in mice', implying by omission that the results are directly applicable to the human health context. Scientist Ben Heathers set up a Twitter account, @justsaysinmice, to highlight this tactic and encourage more responsible headlines that acknowledge where research results apply to mice rather than to humans.

It isn't just mice. Dramatic media headlines often also accompany modelled results, implying that a study directly reflects the real world. As part of the move to escape from Model Land, perhaps it would be helpful for a similar campaign to encourage the addition of 'in models' to this kind of reporting. Distinguishing between Model Land and real world would reduce the sensationalism of some headlines and would also encourage scientific results to clarify more clearly where or whether they are expected to apply to the real world as well. Say we want to modify a headline like 'Planet-Killer Asteroid Will Hit Earth Next Year' to acknowledge the scientific uncertainty. Instead of changing it to '*Models Predict That* Planet-Killer Asteroid Will Hit Earth Next Year', which leaves the reader no wiser about whether it will actually happen or not, we could change it to '*Astronomers Predict That* Planet-Killer Asteroid Will Hit Earth Next Year'. If they turn out to be wrong, at least there is a named person rather than a faceless model there to explain the mistake. In order to have this kind of accountability, we need not only the model and its forecast, but also someone to take responsibility for the step out of Model Land into the real world, which can only be done through expert judgement. Then the question of adequacy-for-purpose does not solely relate to the model, but also to the expert. Are they also adequate for purpose? How do we know?

Past performance is no guarantee of future success

The question that we are starting to address by thinking about adequacy-for-purpose, and which we will explore more in the next chapter, is the extent to which a model can give us reliable, meaningful information about the real world. How far is the model really behaving like the real world, in the ways that matter to us? Which ways *are* the ways that matter? Whether it is a hydraulic model of the economy or a mouse model of human disease, the model can only be evaluated after defining our real-world priorities.

But it is easy to slide over these questions of model evaluation without really addressing them. The language of model 'verification' and 'confirmation' – used in various fields of economics, computing, business and science – implies that a model can be verified or confirmed to be correct, when in most cases it is a simple category error to treat models as something that can be true or false. An exception might be the model that fair six-sided dice have a probability of exactly one-sixth of landing with any face uppermost. This is true, although only in the sense that we have used it to define what we mean by 'fair'. If we measure and find that our dice do not conform to the expectations, we do not say that the model is wrong; instead we call the dice unfair. In effect, the cubical object with numbers on is a model of the idealised situation in our head, rather than the other way around.

As statistician George Box famously said, 'All models are wrong.' In other words, we will always be able to find ways in which models differ from reality, precisely because they are not reality. We can invalidate, disconfirm or falsify *any* model by looking for these differences. Because of this, models cannot act as simple hypotheses about the way in which the true system works, to be accepted or rejected. Phillips would not have argued that his hydraulic model of the British economy was 'true' or 'false', only that it provided a helpful scaffold for thinking, pursuing the consequences of assumptions and seeing the relations of different parts of the economy from a new perspective.

Methods of mathematical inference from models, by contrast, typically do assign some kind of truth value to the models. Statistical methods may assume that model outcomes are related to the truth in some consistent and discoverable way, such as a random error that will converge on zero given enough data, or a systematic bias that can be estimated and corrected. Other statistical methods assume that from a set of candidate models one can identify a single best model or construct a new best model by weighting the candidates according to agreement with other data. Essentially, they assume that the observations are generated by a process that has rules, that those rules can be written formally, that they are sufficiently close to the candidate set of models we are examining and that the only limit to our discovery of the rules is our observation of further data. In Model Land, these mathematically convenient assumptions are genuinely true, but in the real world they can only ever be false.

Box's aphorism has a second part: 'All models are wrong, *but some are useful*.' Even if we take away any philosophical or mathematical justification, we can of course still observe that many models make useful predictions, which can be used to inform actions in the real world with positive outcomes. Rather than claiming, however, that this gives them some truth value, it may be more appropriate to make the lesser claim that a model has been consistent with observation or adequate for a given purpose. Within the range of the available data, we can assess the substance of this claim and estimate the likelihood of further data points also being consistent. Models that are essentially interpolatory, where the observations do not stray much outside the range of the data used to generate the models, can do extremely well by these methods.

The extrapolatory question, of the extent to which it will *continue* to be consistent with observation outside the range of the available data, is entirely reliant on the subjective judgement of the modeller. Depending on the type of model, we may have to ask questions like:

- What kinds of behaviour could lead to another financial crisis, and under what circumstances might they happen?
- Will the representations of sea ice behaviour in our climate models still be effective representations in a 2°C-warmer world?
- What spontaneous changes to social behaviour will occur in the wake of a pandemic?
- Will a species be able to adapt to a changing environment by changing its diet or behaviour after its normal food becomes unavailable?
- Will a certain technology be able to scale up dramatically and remain economically viable without government subsidy?
- What will influence consumer demand for a certain item over the next decade?
- Will there be significant political change?
- Will the 'laws of physics' continue to hold?

These are questions that cannot be answered either solely in Model Land or solely by observation (until after the fact): they require a judgement about the relation of a model with the real world, in a situation that has not yet come to pass. On some there may be general agreement and on others a wide variety of opinion. Answering these questions in different ways could result in completely different, perhaps even contradictory statements of uncertainty, *any or all of which may be valid within the sphere of the model itself*. Or different answers may result in completely different models being applicable to the situation and different forecasts being made. Yet, as there is only one real world, two contradictory forecasts cannot both be correct.

Model diversity

If there are many opinions, many perspectives, many differing value judgements about what the purpose of the exercise is, then there will be many possible models. Sometimes different models may arrive at similar answers, which might give us confidence in those answers. In other cases, their predictions may differ greatly, and the quality of the predictions cannot be distinguished solely by looking at past performance.

Diversity in boardrooms is shown to result in better decision-making by businesses: a more diverse board is able to consider different perspectives, act more imaginatively, challenge each other and improve business performance. The argument for model diversity is extremely similar. Genuine model diversity will help us to consider different perspectives: what really is the bottom line? What is it that we are trying to achieve? It will help us to act more imaginatively: are we being constrained into certain courses of action because we only have one model, which can only describe certain types of interventions? It will allow us to challenge the assumptions of each model, instead of having a single straw man to accept or reject. And it will improve the performance of any decision-making system.

As should be clear by now, this requires not just plurality of models or a token diversity generated by small incremental changes to existing models. It requires genuine diversity of the underlying principles. This might mean creating a climate model that begins with resource consumption patterns rather than with atmospheric fluids, or an epidemiological model that begins with the mental health of the individual rather than the mortality statistics of a population. What would those look like? I have no idea, but bearing in mind the vast resources available to the 'standard' models, there would be little to lose and a lot to gain by exploring some different avenues. Later, we will explore some of the ways in which that might be done.

*

When making a model, we do not often examine the starting point. Whether you are using a computer because it happens to be the thing in front of you, or a sack of plumbing because you haven't got access to a computer, or a mouse because that is what generations of previous PhD students have used, it's important to consider the ways in which the tools and materials to hand might be constraining your imagination. What is the real-world question that you want to answer? What are the kinds of real-world actions you want to be able to take?

If we want to ensure adequacy for a certain real-world purpose, we must ensure that we enter into the right part of Model Land, one that has a view of the part of the real world we are interested in. We can get better and more reliable information by triangulating the views from different parts of Model Land, but we do need to know that our telescopes are fixed on the real world.

3
Models as Metaphors

'Have you guessed the riddle yet?' the Hatter said, turning to Alice again.

'No, I give it up,' Alice replied: 'What's the answer?'

'I haven't the slightest idea,' said the Hatter.

Lewis Carroll, *Alice's Adventures in Wonderland* (1865)

Why is a raven like a writing-desk?

Lewis Carroll had no particular answer in mind to the Mad Hatter's riddle – 'Why is a raven like a writing-desk?' – when he wrote *Alice's Adventures in Wonderland*, but it has vexed readers for years. Many have come up with their own answers, such as 'One is good for writing books and the other for biting rooks.'

Presented with any two objects or concepts, more or less randomly chosen, the human mind is remarkably good at coming up with ways to identify the similarities between them, despite all the other ways in which they might differ. Internet 'memes', for example, are small instances of shared metaphors which rely on pre-existing structures (the picture, which creates a framework story) to be loaded with a new meaning by the overlaid text which identifies something else as being the subject of the metaphor. To take an example that has been around for a long time: the image of Boromir (Sean Bean) from Peter Jackson's 2001 film *The Fellowship of the Ring* saying, 'One does not simply walk into Mordor.' This meme has been repurposed many times to imply that someone is failing to take a difficult task seriously. Other memes include 'Distracted Boyfriend' (one thing superseded by another), 'Picard Facepalm' (you did something predictably silly), 'Running Away Balloon' (being held back from achieving an objective) and 'This Is Fine' (denial of a bad situation). This capacity for metaphor, and elaboration of the metaphor to generate insight or amusement, is what underlies our propensity for model-building. When you create a metaphor, or model, or meme, you are reframing a situation from a new perspective, emphasising one aspect of it and playing down others.

Why is a computational model akin to the Earth's climate? What does a Jane Austen novel have to tell us about human relationships today? In what respects is an Ordnance Survey map like the topography of the Lake District? In what way does a Picasso painting resemble its subject? How is a dynamic-stochastic general equilibrium model like the economy?

These are all models, all useful and at the same time all fallible and all limited. If we rely on Jane Austen alone to inform our dating habits in the twenty-first century, we may be as surprised by the outcome as if we use an Ordnance Survey map to attempt to paint a picture of Scafell Pike or a dynamic-stochastic general equilibrium model to predict a financial crisis. In some ways these models can be useful; in other ways they may be completely uninformative; in yet other ways they could be dangerously misleading. What does it mean, then, to make a model? Why is a raven like a writing-desk?

Creating new metaphors

In one view, the act of modelling is just an assertion that A (a thing we are interested in) is like B (something else of which we already have some kind of understanding). This is exactly the process of creating a metaphor: 'you are a gem'; 'he is a thorn in my side'; 'all the world's a stage', etc. The choice of B is essentially unconstrained – we could even imagine a random-metaphor generator which chooses completely randomly from a list of objects or concepts. Then, having generated some sentence like 'The British economy in 2030 will be a filament lightbulb' or 'My present housing situation is an organic avocado', we could start to think about the ways in which that might be insightful, an aid to thinking, amusing or totally useless.

Even though B might be random, A is not: there is little interest in the metaphor sparked by a generic random sentence 'A filament lightbulb is an organic avocado'. We choose A precisely because we are interested in understanding its qualities further, not by measuring it or observing it more closely, but simply by rethinking its nature by reframing it in our own mind and reimagining its relationship with other entities.

This is closely linked to artistic metaphor. Why does a portrait represent a certain person or a painting a particular landscape? If for some viewers there is immediate visual recognition, for others it may be only a title or caption that assigns this meaning. Furthermore, some meanings may be assigned by the beholder that were not intended by the artist. The process of rethinking and reimagining is conceived by the artist (modeller), stimulated by the art (the model) and ultimately interpreted by an audience (which may include the artist or modeller).

In brief, I think that the question of what models are, how they should be interpreted and what they can do for us actually has a lot to do with the question of what art is, how it can be interpreted and what it can do for us.

Of course, just as some artworks are directly and recognisably representative of their subject, so are some models. Where we have a sufficient quantity of data, we can compare the outputs of the model directly with observation and thus state whether our model is lifelike or not. But this is still limited: even though we may be able to say that a photo is a very good two-dimensional representation of my father's face, it does not look a great deal like him from the back, nor does it have any opinions about my lifestyle choices or any preferences about what to have for dinner. Even restricted to the two-dimensional appearance, are we most interested in the outline and contours of the face, or in the accurate representation of colour, or in the portrayal of the spirit or personality of the individual? That probably depends on whether it is a photograph for a family album, a passport photo or a submission for a photography competition.

So the question 'how good is B as a model for A' has lots of answers. In ideal circumstances, we really can narrow A down to one numerical scale with a well-defined, observable, correct answer, such as the daily depth of water in my rain gauge to the nearest millimetre. Then my model based on tomorrow's weather forecast will sometimes get the right answer and sometimes the wrong answer. It's clear that a right answer is a right answer, but how should I deal with the wrong answers?

- Should one wrong answer invalidate the model despite 364 correct days, since the model has been shown to be 'false'?
- Or, if it is right 364 days out of 365, should we say that it is 'nearly perfect'?
- Does it matter if the day that was wrong was a normal day or if it contained a freak hurricane of a kind that had never before struck the area where I live?
- Does it matter whether the day that it was wrong it was only out by 1mm, or it forecast zero rain when 30mm fell?

As with the example of the photograph, the question of how good the model is will have different answers for the same model depending on what I want to do with it. If I am a gardener, then I am probably most interested in the overall amount of rain over a period of time, but the exact quantity on any given day is unimportant. If I am attempting to set a world record for outdoor origami, then the difference between 0mm and 1mm is critical, but I don't care about the difference between 3mm and 20mm. If I am interested in planning and implementing flood defence strategies for the region, I may be uninterested in almost all forecasts except those where extremely high rainfall was predicted (regardless of whether it was observed) and those where extremely high rainfall was observed (regardless of whether it was predicted).

There are many different approaches to understanding and quantifying the performance of a model, and we will return to these later.

For now, it's enough to note that the evaluation of a model's performance is not a property solely of the model and the data – it is always dependent on the purpose to which we wish to put it.

Taking a model literally is not taking a model seriously

To treat the works of Jane Austen as if they reflected the literal truth of goings-on in English society in the eighteenth century would be not to take her seriously. If the novels were simple statements about who did what in a particular situation, they would not have the universality or broader 'truth' that readers find in her works and which make them worthy of returning to as social commentary still relevant today. Models can be both right, in the sense of expressing a way of thinking about a situation that can generate insight, and at the same time wrong – factually incorrect. Atoms do not consist of little balls orbiting a nucleus, and yet it can be helpful to imagine that they do. Viruses do not jump randomly between people at a party, but it may be useful to think of them doing just that. The wave and particle duality of light even provides an example where we can perfectly seriously hold two contradictory models in our head at once, each of which expresses some useful and predictive characteristics of 'the photon'.

Nobel Prize-winning economist Peter Diamond said in his Nobel lecture that 'to me, taking a model literally is not taking a model seriously'. There are different ways to avoid taking models literally. We do not take either wave or particle theories of light literally, but we do take them both seriously. In economics, some use is made of what are called stylised facts: general principles that are known not to be true in detail but that describe some underlying observed or expected regularity. Examples of stylised fact are 'per-capita economic output (generally) grows over time', or 'people who go to university (generally) earn more', or 'in the UK it is (generally) warmer in May than in November'. These stylised facts do not purport to be explanations or to suggest causation, only correlation.

Stylised facts are perhaps most like cartoons or caricatures, where some recognisable feature is overemphasised beyond the lifelike in order to capture some unique aspect that differentiates one particular individual from most others. Political cartoonists pick out prominent features such as ears (Barack Obama) or hair (Donald Trump) to construct a grotesque but immediately recognisable caricature of the subject. These features can even become detached from the individual to take on a symbolic life of their own or to animate other objects with the same persona.

We might think of models as being caricatures in the same sense. Inevitably, they emphasise the importance of certain kinds of feature – perhaps those that are the most superficially recognisable – and ignore others entirely. A one-dimensional energy balance model of the atmosphere is a kind of caricature sketch of the complexity of the real atmosphere. Yet despite its evident oversimplification it can nevertheless generate certain useful insights and form a good starting point for discussion about what additional complexity can offer. Reducing something or someone to a caricature, however, can be done in different ways – as the most-loved and most-hated political figures can certainly attest. It is effectively a form of stereotype and calls to the same kind of psychological need for simplicity, while introducing subjective elements that draw upon deep shared perspectives of the author and their intended audience.

Thinking of models as caricatures helps us to understand how they both generate and help to illustrate, communicate and share insights. Thinking of models as stereotypes hints at the more negative aspects of this dynamic: in constructing this stereotype, what implicit value judgements are being made? Is my one-dimensional energy balance model of the atmosphere really 'just' a model of the atmosphere, or is it (in this context) also asserting and reinforcing the primacy of the mathematical and physical sciences as the only meaningful point of reference for climate change? In the past I have presented it myself as 'the simplest possible model of climate change', but I am increasingly concerned that such framings can be deeply counterproductive. What other 'simplest possible models' might I imagine, if I had a mind that was less encumbered by the mathematical language and hierarchical nature of the physical sciences? Who might recognise and run with the insights that those models could generate?

Nigerian writer Chimamanda Ngozi Adichie points out 'the danger of the single story': if you have only one story, then you are at risk of your thinking becoming trapped in a stereotype. A single model is a single story about the world, although it might be written in equations instead of in flowery prose. Adichie says: 'It is impossible to talk about the single story without talking about power ... How they are told, who tells them, when they're told, how many stories are told, are really dependent on power.'

Since each model represents only one perspective, there are infinitely many different models for any given situation. In the very simplest of mathematically ideal cases of the kind you might find on a high-school mathematics exam, we could all agree that *force equals mass times acceleration* is the 'correct model' to use to solve the question of when the truck will reach 60mph, but in the real world there are always secondary considerations about wind speed, the shape of the truck, the legal speed restrictions, the state of mind of the driver, the age of the tyres and so on. You might be thinking that clearly there is one best model to use here, and the other considerations about wind speed and so on are only secondary; in short, any reasonable person would agree with you about the first-order model. In doing so, you would be defining 'reasonable people' as those who agree with you and taking a rather Platonic view that the mathematically ideal situation is closer to the 'truth' than the messy reality (in which, by the way, a speed limiter kicked in at 55mph). This is close to an idea that historians of science Lorraine Daston and Peter Galison have outlined, that objectivity, as a social expectation of scientific practice, essentially just means conforming to a certain current set of community standards. When the models fail, you will say that it was a special situation that couldn't have been predicted, but *force equals mass times acceleration* itself is only correct in a very special set of circumstances. Nancy Cartwright, an American philosopher of science, says that the laws of physics, when they apply, apply only *ceteris paribus* – with all other things remaining equal. In

Model Land, we can make sure that happens. In the real world, though, the *ceteris* very rarely stay *paribus*.

Your next objection is probably that we can test our models against data and so settle the question directly regarding which is best. But as I described earlier, the evaluation of a model's performance is always dependent on the purpose to which we wish to put it. If we can agree on the purpose and find a metric that is directly related to that purpose, then we can decide which model has been best. But we also need to agree on what degree of accuracy is sufficient. Let's say we model the shape of the truck in detail and find that it only makes a small difference, then we can decide that the extra complexity is not justified by the improvement in the answer. In order to do that, we would also need to agree that the difference is not problematic: for instance, if we want to decide which truck to use for commercial deliveries, half a second doesn't matter, but if we are planning a movie stunt in which a motorbike crosses the path of the truck, half a second could be critically important and the more complex model justified.

Baking a cake without a recipe

Historian and philosopher of science Marcel Boumans said that 'model-building is like baking a cake without a recipe'. We have a general shared idea of what the outcome should look like, the kinds of ingredients we are likely to need and the types of operations (mixing, heating) that will get us to the point of having a product. Very different kinds of ingredients and operations can have equally delicious outcomes, all of which we would recognise as cake. Conversely, even the best ingredients and superficially reasonable operations can result in an inedible mess if combined incorrectly, a disappointment many home bakers will be familiar with. Using a recipe, we can (almost) guarantee a good outcome, but making up a recipe from scratch is surprisingly difficult. After I lost my favourite flapjack recipe, it took several iterations of experimentally combining what I thought I could remember of the ingredients to find a balance between crumbly granola and tooth-destroyingly solid lumps. In retrospect, I should definitely have just searched for another recipe or left it to my husband, who is a considerably better baker than I am.

Like baking without a recipe, some people are also better at modelling than others and are able to say from the outset that one kind of strategy is likely to work and another likely to fail, to know that some combination of ingredients typically works well, or how long to bake the cake to achieve the ideal result. There is a definite art to balancing the different demands effectively, making judicious assumptions, and deciding when to continue making improvements and when to stop and avoid overfitting.

Recipe or no, the proof of the pudding is in the eating. In the era of vegan, gluten-free and raw cakes which may share no ingredients or methods with a traditional British Victoria sponge, perhaps there is a lesson to be learned that we should be less attached to traditional methods and more focused on achieving positive outcomes. Taking an adequacy-for-purpose approach, if you could reasonably put candles on it and sing 'Happy Birthday', then it probably counts as a cake, where just defining it by the ingredients or methods used wouldn't necessarily get you anything presentable. But adequacy-for-purpose alone can also be problematic when we risk dressing something up as something it isn't – in general, it is helpful to have some indication of what is inside. You could put candles on a round of cheese and sing 'Happy Birthday', but the guests might be a little surprised when they bit into it.

This analogy also highlights the culturally specific aspects of model-building. As I am British, I have expectations of what 'cake' means that come from my own upbringing and cultural surroundings. I know what *I* mean by a cake! But American readers may have a slightly different idea; German readers a different idea again; Indonesian readers a very different idea – and the notion of 'putting candles on it and singing "Happy Birthday"' might also need to be recast into other settings. Similarly, a model that is produced in a certain cultural context, such as a British tech company consisting mainly of Oxbridge engineering graduates, would draw on the mathematical and social traditions of that environment both for the conceptual and the technical construction of the model and the definition of adequacy-for-purpose. And this often happens without the makers even realising that any other perspectives exist.

Clever horses and stochastic parrots

Clever Hans was a performance act at the beginning of the twentieth century, a horse that was able to tap a hoof to indicate the correct answer to basic arithmetic questions. The show was very convincing, but further investigation revealed that the horse was responding to subtle changes in body language in his handler and could only give the right answer if his handler knew it.

When models become more like black boxes, some may become what has been termed a 'horse': a system that is not actually addressing the problem it appears to be solving. Examples of such horses are common in machine learning, where trial data can become accidentally confounded with other conditions that the machine learns. One probably apocryphal example is a 'tank detector' able to identify camouflaged military vehicles in photographs with remarkable accuracy. The story goes that, in both the training set of images and the test set, the presence of tanks was strongly correlated with sky/weather conditions and the algorithm had simply learned to distinguish cloudy days from clear skies. Clearly that kind of skill would not persist outside the development environment. But other correlations are more insidious. Researchers have shown how image-recognition models can be heavily biased with social stereotypes, consistently mislabelling pictures of men cooking as women, for example. Automatic language translation models will take a gender-neutral sentence in one language – 'They are a mechanic', 'They are a nurse' – and make it a gendered sentence in another language – 'He is a mechanic', 'She is a nurse'. You can see why this might happen, but is it a good thing because it makes the classifier more accurate, or is it a bad thing because it reinforces stereotypes? The assessment of adequacy-for-purpose depends on the purpose you think these models are working towards. A purely statistical metric will score better by using these correlations, but it may have corrosive social effects.

Cathy O'Neil has written about the use of recidivism models in criminal sentencing, describing how someone would be assessed by a model using information about their employment status, neighbours and family. If the model deems them to be a higher reoffending risk, they get sentenced to a longer incarceration. As O'Neil puts it:

> He is finally released into the same poor neighborhood, this time with a criminal record, which makes it that much harder to find a job. If he commits another crime, the recidivism model can claim another success. But in fact the model itself contributes to a toxic cycle and helps to sustain it.

There are also blind spots. Google Photos' image-labelling in 2015 disturbingly managed to classify pictures of Black people as gorillas, a mistake that was unsatisfactorily corrected by deleting the label 'gorilla'. This meant that a picture of a gorilla could not be recognised at all, preventing a particular mistake being made again but totally failing to address the question of why it had occurred in the first place. In 2020, a video uploaded to Facebook by a British tabloid newspaper including clips of Black men was automatically labelled with the category 'primates'. These offensively inadequate models are then incorporated into even more consequential applications. Whether it is a mobile phone that cannot detect the face of the user or self-driving cars that are unable to recognise a human pedestrian, there are many reasons to be appalled by the continued appearance of this kind of mistake in the context of the rapid development of real-world artificially 'intelligent' autonomous systems. With larger and larger data sets and aspirations to more widespread use, the prospect of only removing this kind of error by hand after the event is infeasible and unacceptable. Use of such systems presents a real-world harm to particular individuals and communities.

So these questions of social and cultural bias are something with which the artificial intelligence community has to grapple. But it's difficult, because most of the researchers are WEIRD and educated at elite schools that emphasise a dominant paradigm of science where a model's accuracy is the highest virtue. Questions of power, bias and implications for marginalised communities do not arise naturally because they do not personally affect the majority of these researchers.

Linguist Emily Bender, AI researcher Timnit Gebru and colleagues have written about 'the dangers of stochastic parrots', referring to language models that can emulate English text in a variety of styles. Such models can now write poetry, answer questions, compose articles and hold conversations. They do this by scraping a huge archive of text produced by humans – basically most of the content of the internet plus a lot of books, probably with obviously offensive words removed – and creating statistical models that link one word with the probability of the next word given a context. And they do it remarkably well, to the extent that it is occasionally difficult to tell whether text has been composed by a human or by a language model. Bender, Gebru and colleagues point out some of the problems with this. English-language text on the internet is not from a random sample of individuals but is dominated by those with time and inclination to write in that medium. And removing text containing potentially offensive words can also result in removing text produced *by* minority communities. As such, they note that 'white supremacist and misogynistic, ageist, etc. views are overrepresented in the training data, not only exceeding their prevalence in the general population but also setting up models trained on these datasets to further amplify biases and harms'. Again, this is a source of real-world harm to individuals, and not just something in Model Land.

The key distinction between a 'horse' and a model that is simply wrong is that, on some subset of data, the 'horse' *does* get the right answer and therefore its problems pass undetected. If we are only assessing the performance of the 'horse' with respect to that data, we will have to conclude that it is a very good model – or, for example, that Clever Hans is in fact able to perform arithmetic. The detection of the confounding factors in the case of an unexpectedly mathematically talented equine happened because of an underlying expectation (or prior belief, if you are a subscriber to the Bayesian formalism of statistically updating probabilities, which I will describe in more detail at the end of chapter 4) that a horse could not possibly do those things and that a different explanation for the observations should therefore be sought. With the tank-detection story, it was similar scepticism that prompted investigation. This is the reason why it is so important to ensure that test data are as different as possible from training data, to remove the potential confounding factors that make it look like the model is doing well when it is not. Of course, both Clever Hans and the tank detector were doing interesting things – just not what their handlers thought they were doing. The 'stochastic parrot' language models are doing very interesting things too, and their output may be indistinguishable from human-generated text in some circumstances, but that by itself is not sufficient to justify their use.

Explainability and 'Just So' stories

Apart from finding as much different test data as possible, how can we spot our model horses before they produce horse manure? A concept that has been gaining in importance is *explainability*: to be sure that we have the right answer for the right reason, we want to know the reason for the answer. In other words, we want to be able to give a narrative of how the model arrived at its outcomes. That might be an explanation that the tank detector is looking for edges, or a certain pattern of sky, or a gun turret. It might be an explanation that a criminal-sentencing algorithm looks at previous similar cases and takes a statistical average. If we cannot explain, then we don't know whether we are getting a right answer for 'the right reasons' or whether we are actually detecting sunny days instead of tanks, racial bias instead of – well – is there a way to predict future criminality without resorting to stereotypes? Doesn't it always negate the individual's agency and ability to make life choices? The need for algorithmic explainability and the relation with fairness and accountability, described by Cathy O'Neil in *Weapons of Math Destruction*, is now acknowledged as being of critical importance for any decision-making structures.

I want to extend this thought to more complex models like climate and economic models, and show that, in these contexts, the value of explainability is not nearly so clear cut. For instance, let's say we have a climate model that shows an increase in extratropical cyclone intensity in the course of the twenty-first century. That seems plausible, says one scientist, because it is consistent with our expectations of higher sea-surface temperatures and a moister atmosphere, which feed cyclone activity. But a second climate model shows a decrease in extratropical cyclone intensity over the same period. That seems plausible, says a second scientist, because it is consistent with our expectations of decreased pole-to-equator temperature gradients, inhibiting cyclone activity. What I am highlighting is that explainability by itself is not enough to distinguish a good model from a poor model, especially when there are competing factors working in opposite directions. Even if we might have a reasonable level of intuition about the importance of each factor, pitting them against each other and saying whether the resulting outcome will tip in one direction or the other is a much more difficult task.

Having said that, of course, explainability is immensely attractive because it helps us have confidence that we truly know what we are doing. Model results are always accompanied by explanations that help us to understand why the model is doing what it is doing. These can be both Model Land explanations ('unemployment is high in this model because of X') and bird's-eye-view explanations ('the model does not well represent the drivers of youth unemployment because of Y'). Underlying this is an assumption that if we know what the model is doing, we can also understand whether it is any good, whether it is likely to be any good in the future, and what kinds of Model Land insights we can transfer with confidence into the real world.

That is all well and good. But if we are relying upon explainability as a warrant of confidence, it begs a methodological question: how do we know that we are truly explaining and not just constructing a post-hoc rationalisation? We might call this a 'Just So' story, after Rudyard Kipling's children's tales. One explains 'How the Elephant Got His Trunk' with a story about a crocodile pulling an elephant's nose. Of course, this is an explanation of a sort, in the sense that within the logic of the story (model) it is all plausible, offering both physical mechanism and motivation. Outside the story, however, this 'explanation' is completely in conflict both with observations (like the degree of flexibility of skin) and concepts (such as evolutionary theory), and we can easily agree it to be false, if a nice visual story. Do we know that we always have more direct intuition about a situation that we try to describe than a five-year-old does about the evolution of animal characteristics?

'Just So' stories are constructed both within and about models, and they overlap seamlessly with 'genuine' explanations. In the example of cyclone activity mentioned above, we are constructing a rationalisation rather than a true explanation of why the cyclone activity is observed to increase or decrease in the model. So which one is the true explanation? Well, neither. Both are simplifications; in fact, both mechanisms contribute to the change in cyclone activity, and there are other contributing factors too. If, as I said above, explainability helps us have confidence that we truly know what we are doing, it is still somewhat limited and potentially fallible.

Economic sociologist Ekaterina Svetlova has studied the models used by financial traders and their use of explanation and rationalisation to go either with or against the model. For example, if a model is used to look at past data, project into the future and come up with a statement that 'emerging markets are overvalued', this is not the end of the analysis. The next questions are whether the history is relevant, whether the model contains the right information and so on. The secondary explanation, which comes on top of the model, can either confirm or reject the findings of the model and so support any course of action, almost regardless of what the model says. This kind of post-hoc rationalisation presents an immense problem for statistical analysis, because it enters into the evaluation of model quality or model performance in an unquantifiable way. How far can you stretch a metaphor before deciding that it breaks? If a raven is like a writing-desk in a certain way, can we extend that to deduce any of its other qualities, or not?

In the absence of an answer provided by the Hatter himself, it does not make sense to say that 'One is a pest for wrens and the other is a rest for pens' is a better answer than 'Because outstanding bills are found on both of them'. The only arbiter here is our personal sense of humour. How is a model like the real world? Again, there are many different ways to associate a model with something real. We are fortunate that, for *some* of these possible associations, we can make a quantitative assessment of the degree of similarity. But, in many other important respects, no quantitative assessment may be possible: the quality of explanation; the realism of key behaviours; the communicability of insight; the framing of a certain perspective; the insights generated by pondering the metaphor. Although these insights cannot be expressed in numbers, they are no less valuable.

Whose cake?

I have argued in this chapter that we should be thinking about our models as metaphors. Just as there are many metaphors for any situation, so there are many possible models. The value of metaphor is in framing and reframing the situation so that we can see it from a new perspective, make unexpected links, and create stories and explanations that help us to think collectively as well as individually about the implications of the information we have. Models have all of those qualities as well, and in addition they can sometimes generate reliable quantitative forecasts of some measurable outcome.

Because metaphors rely on shared experiences, their construction and interpretation is also dependent on social context. When you bake a cake without a recipe, you draw upon years of experience, maybe of baking, maybe of seeing other people baking, maybe of eating cake or seeing pictures in recipe books and shop windows – all things that will be different according to who and where you are.

Being able to explain the process, and also to explain the cultural significance and purpose of a particular model, will help others to assess the degree to which it is adequate for purpose. Context is important. A cake made by a four-year-old for his grandmother can be wonderfully suited for purpose even if it fails to meet objective metrics of competence, because it is not only a food item but also an expression of love. An image classifier that labels some people as animals may be the 'best available' model in statistical terms, but it has completely unacceptable potential consequences.

Quantitative accuracy is only one part of the competence of a model. If we only ever assess mathematical models by their quantitative accuracy, we are ignoring their social and cultural context. If mathematical models are not explicitly designed to express care for their subjects, they entrench a lack of care.

So we need to construct our models as metaphors carefully. When making either conceptual or mathematical decisions about the structure of our models, we are choosing in which directions to extend our imagination. Taking this perspective, I think, helps us to understand both why 'all models are wrong' and also why it is often unhelpful to look for the 'best' model by mathematically testing candidates against certain objectives. Each model provides a different view and embodies a different idea about the important characteristics of the thing we are modelling.

4
The Cat that Looks Most Like a Dog

> Not only was it difficult for him to comprehend that the generic symbol *dog* embraces so many unlike individuals of diverse size and form; it bothered him that the dog at three fourteen (seen from the side) should have the same name as the dog at three fifteen (seen from the front).
>
> J. L. Borges, 'Funes the Memorious' (1954)

The mathematical methods used to analyse models are as important as the models themselves. We need expert judgement to make models – which gets us into Model Land – and we also need expert judgement to get *out* of Model Land, through analysis and interpretation of model results. When making forecasts about the future, the mathematical probabilities we come up with depend on those expert judgements. As Nassim Taleb has said, 'probability is a qualitative subject'. In this chapter, we will look at why that is the case, and how different mathematical assumptions about models might lead us to come to different conclusions about what those models tell us.

The reference class problem

The fundamental problem of trying to model the real world is that real life only happens once: every situation is different. The question that we are asking when we try to do anything we might describe as 'science' is: which A are sufficiently like B to be a useful analogue? When we make that analogy – or metaphor, as I described it before – and use it to make inferences about the (unknown) properties of B given the (observable) properties of A, we are making a claim about the degree of useful similarity.

Members of a very limited class of real-world objects are pretty close to being Model Land objects, mathematical idealisations:

- Which dice are sufficiently like my dice to give a good estimate of how often I could expect to throw a six?
- Which coins are sufficiently like my coin to give a good estimate of how often I might flip it and get twelve heads in a row?
- Which electrons are sufficiently like my electron to give a good estimate of how it will behave in a magnetic field?

Instead of throwing thousands of other dice or flipping thousands of coins to get a reproducible answer, we are often pretty sure that we can use a mathematical idealisation of Ideal Dice (with a probability of exactly one-sixth of landing on any face) or Ideal Coins (with a probability of exactly one-half of landing on either heads or tails). No doubt, though, our confidence is derived from the observations of many millions of past dice-throwers and coin-tossers, as well as a belief in the symmetry of the situation.

Most real-world objects, by contrast, are not close to being mathematical idealisations. We might, however, have a large amount of relevant data about other, similar objects. In these cases, we resort to statistics of things that can be observed to infer the properties of the one that cannot be observed or that has not happened yet:

- Which people are sufficiently like me to give a good estimate of my risk of death if I contract influenza?
- Which people are sufficiently like me to give a good estimate of my risk of death if I contract Covid-19?
- Which bicycles are sufficiently like my bicycle to give a good estimate of how many more miles it will go before it needs a new chain?
- Which ladders are sufficiently like my ladder to give a good estimate of how much weight it can bear without breaking?

- Which days are sufficiently like tomorrow to give a good estimate of tomorrow's weather?

For the first question (influenza), there is a lot of data available. I can break down the sample by age, sex, country of residence, socioeconomic status, smokers/non-smokers and many other relevant or potentially relevant variables. The greater the number of variables I choose to sort by, the fewer the number of people in the sample that I can use to estimate my personal risk. But of course what I hope to do by this is increase the *relevance* of that sample to me, personally. I have to choose variables that I think will have some bearing on this particular risk – it is unlikely, for instance, that my favourite colour is relevant. Note, though, that I don't want to increase the relevance indefinitely: I still want to have enough people in the sample so that I get a statistically meaningful answer. To take it to absurdity, I could say that I myself have had influenza in the past and did not die, therefore my risk of death if I were to contract it again is exactly zero (zero deaths divided by one instance). This is wrong because many things about me, including risk factors such as my age, have changed since I last had it. So I may not even be a good analogy for myself.

For the second question (Covid-19), at the time of writing I have not contracted this disease. But there are also a (growing) number of 'people like me' who have contracted it, and therefore I can estimate my risk from their experiences in exactly the same way as the influenza risk. Due to the smaller numbers, I would have to accept either less relevance (e.g., taking into account only age and sex rather than other risk factors) or greater statistical uncertainty (due to small sample sizes) in the number I come up with.

In both cases, I also have a further degree of uncertainty that is not captured by the statistics: the possibility that the disease I get will not be exactly the same as the one all these other people got. Viruses mutate regularly and the risk of death from one strain may be significantly different from the risk of death from another. I have no way of incorporating this possibility into my statistical analysis other than by looking up from my calculation and understanding it to be incomplete.

In a similar way, we can take a subset of cars (perhaps those of the same make and model and with a similar number of miles on the clock) to estimate the characteristics of my car, a subset of ladders (perhaps a sample taken at the factory by the quality-control engineers and tested to destruction) to estimate the characteristics of my ladder, and a subset of days (perhaps those on the same day of the year or with similar preceding weather conditions) to estimate the characteristics of tomorrow's weather. The key concept here is *exchangeability*: the idea that, although the members of a group may be slightly different, no one member is special and the characteristics of the group are a good statistical estimate of the characteristics of the individual. Then all we have to do is define the group or reference class within which all members are exchangeable.

I am making this sound easy. In practice, it requires a lot of information and expert judgement, and, as described above, there is always a trade-off between having *more* data and having *more relevant* data.

So if dice, coins and electrons are essentially mathematical idealisations already, and members of groups such as cars, ladders and weather days can be estimated statistically, what about everything else? Some things simply do not come from a clear reference class of other, similar things. The Earth is a planet, but we probably cannot draw very many useful conclusions about it through a statistical analysis of other planets. The next US presidential election is one of many elections in US history and across the world, but analysis of previous results cannot possibly be sufficient to predict the next. The next pandemic will come from one of a fairly limited number of families of viruses, but we can draw only limited (though useful) conclusions about it by examining the properties of the others.

So we make models. Models can be physical, mental, mathematical or computational, and they are by nature metaphorical. The reference class question is about how far the metaphor can be extended:

- Are mice sufficiently like humans to make it possible to estimate human drug reactions from mouse drug reactions?

- Are linear regression models sufficiently like human choices to make it possible to estimate the chance that you will buy the pair of shoes advertised by Amazon from my predictive analytics dashboard?
- Are climate models sufficiently like the Earth system to make it possible to estimate real climate sensitivity from modelled climate sensitivity?
- Are cats sufficiently like dogs to make it possible to estimate the chance of Fido being friendly given my experience with Felix? Or to operate on a dog, if I have been trained as a veterinary surgeon for cats?

We also have to ensure that the model represents something real, and in order to do so we generally start by observing and collecting data.

Big Data and Small Data

Big Data are very fashionable. It is easier than ever to collect and store a very large amount of data without even knowing what you are going to do with it.

But this is not really a book about Big Data. This is a book about Small or Not-Directly-Relevant Data and Big Ideas. Statistics is what you do when you have Big Data. Modelling is what you do when you don't have enough data; or when you're not really sure that the data you have are actually the data you want; or when you *know* something about the patterns and structures (Ideas) that are going into the data and you don't want to just call it all random variation.

When I say Small Data, I don't necessarily mean that we only have a handful of data points. In the field of numerical weather and climate prediction, we have an extremely large amount of data, of many different kinds. Relative to the question of deciding what tomorrow's weather might look like, this is Big Data. We can define a class of weather models that are close enough to being exchangeable with tomorrow's weather observations to ensure that we have genuinely, quantifiably good predictions. (I know it may not seem like it to British readers for whom complaining about the weather forecast is something of a national sport: remember, a good prediction may still include significant uncertainty.) Relative to the question of deciding what the weather in the 2080s will be like, however, it is definitely Small Data.

We know from observation that greenhouse gas concentrations in the atmosphere are rising, and we know from first physical principles that this significantly alters the energy balance of the Earth system. We also see statistical evidence that the climatic statistics are indeed changing over periods of decades, and in particular that temperatures are increasing, in patterns that are in accordance with our physical expectations. Taken together, what this means is that the Earth system of today is not the same as the Earth system of fifty years ago, nor the same as the Earth system of fifty years in the future.

In other words, old and present observations of weather are not exchangeable with future observations of weather, so we need to make a climate model that does not just represent the current conditions but aims to project forward into a fundamentally different future. While the predictions made by weather models are effectively interpolatory in nature, the predictions made by climate models are extrapolatory. Instead of using present observations to generate new observations directly, we use the observations to inform our understanding of physical processes, then use those physical processes to generate new observations. Instead of assuming that the distribution of observations is fixed over time, we assume that some underlying generative process remains the same over time (the laws of physics).

Even though we have a very Big quantity of Data to answer questions like this, it is not necessarily relevant data. We have similarly huge amounts of data on the microscale movements of stock markets and exchange rates, but these are still very difficult to predict.

I have distinguished statistics from models in the above to make the point clearer, but they are really two ends of a spectrum rather than being completely different beasts. Models encode statistical assumptions; statistical methods are, of course, models too.

Similarly, Big Data and Small Data are ends of a spectrum: the lower end is Zero Data. We'd better not talk about Actively Misleading Data yet, though I will return to it in chapter 6.

Model quantities are not real quantities

What's the freezing point of water? Easy. Zero degrees Celsius, right? Or 32 Fahrenheit, or 273 Kelvin, depending on your preferred temperature scale. Now, if you're making a model that includes the possibility that water might freeze, you're probably sufficiently sure of that value to hard-code it into the model. If you have salt water or very high pressure, you will probably look up the measured variation of the freezing point and hard-code in those variations. You will also hard-code in other physically accepted values such as the mass of an electron or the acceleration due to gravity. After all, they have measured values. Wouldn't it be crazy to do anything else?

It sounds like a silly question but there is a real issue here. Let me give some examples that sound less ridiculous. First, think about wind speed in a model with 100km-wide grid boxes. The real wind speed can't be measured on a 100km grid, it can only be measured at a point. The model wind speed is not defined at a point; it can only be said to be identical to the value in the whole 100km grid box. So, if we have measured a value of wind speed at some point (maybe even at the exact centre) of the grid box, we can't just put that measured value directly into the model. The process of taking data and reforming them into something that can be meaningfully input into a model is called *data assimilation*. Data assimilation is a very wide field of study, particularly important in disciplines such as weather forecasting, and there are many complex methods for deciding exactly how the observations we make in the real world can be projected into Model Land.

A second example is the viscosity of a fluid. The viscosity of a real fluid is defined as its resistance to shear stresses: effectively a measure of the microscopic internal frictional forces within the fluid. Treacle has a very high viscosity, water has a medium viscosity and the gases in the air (which is also a fluid) have a very low viscosity. So let's say we measure the viscosity of a fluid and then try to model it on a computer using a numerical grid. Unfortunately, something goes wrong, because the numerical grid is much too large to capture the viscous dissipation on microscopic scales. We are simply not modelling the molecular scale, which is where the forces causing viscosity actually happen. There absolutely *is* a viscosity in the system, but if you plug in the measured value you get a picture that looks nothing like the fluid you are trying to model. One solution is to use what is called an 'eddy viscosity', which represents the overall net frictional dissipation on the model scales, and which is a function of the size of the grid boxes in our model as well as the properties of the fluid. The eddy viscosity is then set to a value that makes the simulation look realistic.

The molecular viscosity and eddy viscosity are not close: they have the same units and fulfil the same purpose in an equation, but their values differ by orders of magnitude. Now let's go back to the freezing point of water and the acceleration due to gravity: surely it is also possible that these physical constants might be better represented in a numerical model by values other than the one that is observed by measurement in the real world? If appropriate values were orders of magnitude different, like the viscosity, somebody would probably have already noticed, but what if in a weather forecast model the best value for the freezing point of pure water at standard pressures were half a degree different from the 'true' measured value? As far as a physicist is concerned, the properties of water are not negotiable; but in a model everything is negotiable and every decision implies a trade-off.

If the 'viscosity' of the model-atmosphere is set at the observed value, then everything looks wrong immediately. If the modelled freezing point of water is set at the observed real-world value, possibly it may also limit the ability of the model to make good forecasts. I am not saying we should throw out our physical knowledge; I am saying that when we make a model we are projecting our real-world knowledge into Model Land. Model Land is a space where the only rules that apply are the ones we have created for ourselves; these may or may not coincide with physical laws and relationships in the real world. In Model Land there is genuinely no reason for the freezing point of model-water to be 0°C, except that you say that it is.

When we step into Model Land in this way, we make some of our assumptions absolutely rock solid, while others are variable, to be assigned with reference to data. This is a function of our thinking about the real-world situation: what aspects are most strongly part of our internal conceptualisation of how the world works? Yet those assumptions that are deemed to be rock solid are still assumptions. It would help to distinguish *all* Model Land quantities from their real-world counterparts: this is the model-viscosity; that is the real-viscosity. This is the model-wind speed; that is the real-wind speed. This is the model-gravity; that is the real-gravity.

But of course these assumptions also fulfil an important role in anchoring our model in reality. If a model cannot give the correct answer without dramatically changing a known, observable quantity away from its real-world value, does this not tell us something about the quality of the model?

To take the counterargument to extremes, what about π? What if we could get a better fit to real-world observed data by recalibrating the value of π away from its mathematical-constant value of 3.14159 ... or by breaking the laws of conservation of energy and mass? Would we do that? Would *you* do that?

You have to draw the line somewhere. Even I probably wouldn't change π (although my colleague at the London School of Economics Leonard Smith points out that I *do* change the value of π every time I use a digital computer). And yet what you are doing, in making a model, is exactly that: simplifying reality away from the messy truth towards something that happens to be more tractable. Why is it that it feels totally acceptable to make simplifications or add in empirically derived fudge factors along some dimensions, but to do so along others would be complete sacrilege *even if it resulted in a model that could make demonstrably better predictions?*

Model laws are not real laws

If model quantities are not real quantities, what about model laws? Perhaps we were getting close to that when we discussed π and the conservation of mass. Changing these would be violating the laws of physics, wouldn't it?

I think the most coherent argument here is that we often need to impose as much structure on the model as we can, to represent the areas in which we do genuinely have physical confidence, in order to avoid overfitting. If we are willing to calibrate *everything* with respect to data, then we will end up with a glorified statistical model overfitted to that data rather than something that reflects our expert judgement about the underlying mechanisms involved. If I can't make a reasonable model without requiring that π=4 or without violating conservation of mass, then there must be something seriously wrong with my other assumptions. In effect, we are encoding a very strong assumption (or Bayesian prior) that π really *should* be 3.14159 and mass really *should* be conserved – and we would be willing to trade off almost anything else to make it so.

Working backwards, our other model laws are less rigid. Solving any continuous equation numerically requires that we break it down into discrete components. Representing a large number of individuals is often done on an aggregate level, ignoring the differences between those individuals. These are things about which we intuitively have more flexibility, and where trade-offs are more acceptable. And further down are different parameters, choices to ignore certain characteristics and so on, which are the stuff of model development and debate.

I am not at all sure that this hierarchy is a natural one. Western science certainly places the value of π and the conservation of mass on a pedestal, but it would be interesting to consider a modelling science that could invert that hierarchy and insist upon accurate representation, say, of individual species and ecosystems, or individual households disaggregated by geographical and personal characteristics, with the more general 'laws' being fudged to make the evaluation criteria work out OK. Our cultural frame for mathematical modelling tends to mean that we start with the mathematics and work towards a representation of the world, but it could be the other way around.

After all, who said there were any real laws in the first place? Even the most concrete formulations of natural order are only observationally determined and only statements of our best knowledge at the current time. We cannot be sure that the laws of thermodynamics or even conservation laws will not admit modification in future similar to the modifications made to Newtonian mechanics by Einstein's theory of relativity. The best we can say is that they work, now. Of course, that is good enough for most practical purposes.

In this sense 'real' laws are only model laws themselves. Nancy Cartwright has written in detail about how scientific laws, when they apply, apply only with all other things being equal. No wind resistance, no measurement biases, no confounding interactions with other effects, no cat blundering over the keyboard, etc. That's why the study of science requires laboratories: clean, bounded, sterile places where we can ensure, to the best of our ability, that all other things do remain equal. A small outpost of Model Land. That is also why so many scientific results prove inadequate to describe the real world, and why the social sciences, being inextricable from the real-world context, are so much harder than the 'hard' sciences. And even if you do believe in laws, complex systems do not generally admit decomposition into causes, only contributory factors and statistical effects.

All predictions are conditional

Because we have to make that caveat of all other things remaining equal, all predictions are *conditional* predictions, or what some sciences prefer to call projections. Conditional predictions are only predictions *if* a certain set of conditions are true. Those conditions come in many different varieties.

Some conditions are related to the influences that drive them: a model might predict increasing rates of unemployment conditional on continuation of a particular set of government policies, or decreasing rates conditional on adoption of a different set of policies.

Other conditions are get-out clauses identifying possible but unexpected failure modes: I confidently predict that next year I will have more grey hairs than I do now, conditional on not being hit by a bus or undergoing chemotherapy. And I confidently predict that in thirty years' time global mean temperature will be higher than this year, conditional on there being no major volcanic activity, nuclear war, or implementation of geoengineering by solar radiation management.

Sometimes the conditions are all lumped in together, as when the output of a model is provided directly as a forecast in itself. Then every assumption that the model relies upon is a part of the conditions underlying the prediction: 'X will happen, conditional on this model being adequate for the purpose of predicting X.'

Are probabilistic forecasts cheating?

Of course, many forecasts do not predict directly that X *will* happen. Instead, they might forecast only a probability of X happening: 'a 90% chance of rain on Tuesday in New York City' (conditional on my weather model being adequate for the purpose of predicting rain in this region, and on no asteroid hitting the city overnight). On Tuesday, we find that it does indeed rain for part of the day in New York City. Does that mean my forecast was correct? Even better, is it evidence that my model is a good model? Conversely, if it did not rain on Tuesday, would that be evidence that my forecast was wrong or that my model is a bad model, or is it just one of the 10% of outcomes expected to turn out against the forecast?

Probabilistic forecasters are often accused of somehow having their cake and eating it, by being able to claim success for any outcome and to shrug off failure because 'we expect to be wrong some of the time, and this does not invalidate our model'. Surely that's cheating? If we are predicting a single event, this view is more or less right: making a probabilistic forecast is close to meaningless, because only one event will occur and we can't invalidate the model unless it predicted with certainty that something would occur which did not. In the case of weather forecasting, though, we can make a more effective assessment of whether the forecast probabilities actually match up with real-world outcomes. For instance, we can see whether, when rain is forecast with 90% probability, it happens on 90% of occasions. A forecast that gets the right probabilities is said to be 'reliable'. But just getting the right probabilities is not as much use as you might think. For example, in the course of one year in New York, it rains on approximately 30% of days. So if I make a forecast that says '30% chance of rain' every single day, it will be a *perfectly reliable* forecast – and also perfectly useless, because it will not tell you anything that you could use to change your behaviour. Actually, this *perfectly reliable* forecast tells you less than you could know by looking out of a window to see if there are clouds in the sky, or by checking your calendar to see if it is spring or autumn.

What this means is that we need to be careful about how we evaluate probabilistic forecasts. Forecasts need to be not only reliable but also informative. What we want to know, really, is whether we can make better decisions on the basis of a forecast than we could do without that information. If the forecast says 'it will rain tomorrow', then we have an easy decision to make: to bring an umbrella. But if the forecast says '60% chance of rain', we have to trade off the benefits of it being right and the possible costs of it being wrong; and if we aren't sure whether to believe the forecast, we have somehow to factor that in as well. That sounds like a lot of effort – maybe enough to make you invest in a decent raincoat and just wear that every day instead.

But what if you have more weather-sensitive applications, on which something bigger is riding than the state of your hair after a morning walk? Let's say you are growing spring vegetables for sale, trading wheat futures or assigning water allocations in a drought-stricken area. You might have a pretty clear understanding both of how your business will fare given more or less rain on a certain day, and the actions that you would take on receiving a forecast. Therefore, you can do some quantitative optimisation based on the forecast to decide how to act, given the benefits and costs of all possible outcomes. Not only that: you can track your performance over time, comparing the outcomes of the decisions that you did make based on the forecast information with the outcomes of the decisions that you would have made in the absence of that information. In that way you can even generate a numerical figure of how much the forecast was actually worth to you in monetary terms.

In this kind of data-rich environment, therefore, probabilistic forecasts are not cheating at all. They are incredibly useful summaries of the best available data – and they help us to make better decisions. This kind of probability is used to reflect the expected frequency of different outcomes. Over time, we hope, we can compare that to the observed frequencies in order to build up a picture of how the forecast performs in different circumstances.

Are probabilistic forecasts of single events cheating?

Let's turn to a slightly more difficult question now. What if we are genuinely forecasting a single event? In that case there will be no long-term distribution of outcomes, only a single occurrence.

The US presidential elections have been the subject of quantitative forecasting by bodies such as Nate Silver's FiveThirtyEight, which on 8 November 2016 predicted a 71% chance of Hillary Clinton winning that year's presidential vote based on analysis of polling data, corrected for expected sampling biases. Of course, she did not win, and there followed a lot of discussion about why the predictions had been so wrong, what had happened to the polling, etc. FiveThirtyEight's take on it, predictably, was that a 71% chance of one candidate winning still left a 29% chance of the other winning, with Nate Silver then writing articles like 'Why FiveThirtyEight Gave Trump a Better Chance Than Almost Anyone Else' (published 11 November 2016), arguing that, based on the forecast, the result should not have been as shocking as it was.

Is this a coherent defence of single-event probabilities? There are two questions here. The first is a philosophical one about whether such probabilities actually exist in any meaningful sense, an enquiry that could take you down many different rabbit holes. The second is a much more pragmatic one: can you use this probability to make decisions? The purest kind of decision is a simple bet: if you put $1 on the outcome, what odds would you accept? For instance, on 7 November 2016, what return would you have needed to see to put money on Donald Trump winning the election? I think even for those most dismissive of the possibility, a 1,000:1 bet – meaning that you put down $1, losing it in the event of a Trump loss, but in the event of a Trump win, gain $1,000 – would have seemed like a reasonable gamble (even before the possibility of hedging that against other bets offering inconsistent odds). There never was any 'true' probability of Trump becoming president; the probabilities represented only the degrees of belief of different people and organisations like FiveThirtyEight, the *Washington Post* and so on. And yet it is clear that in some sense both outcomes were possible, so the assignment of a probability for the sake of making quantitative decisions is not unreasonable, as long as we recall that the outcome will be one of the two extremes and is not in itself represented by the probabilistic forecast.

Over time, if the forecaster offers probabilities for a lot of different single events, we can build up a picture of whether they are typically too optimistic, too pessimistic or about right. But we have the reference class problem I described earlier: is good performance concerning UK election probabilities an indication of reliability for US presidential elections? Is good performance concerning baseball statistics an indication of reliability for US presidential elections?

There is no cheating going on here. But there is a lot of scope for disagreement about the appropriate reference class by which the reliability of any given class of single-event forecast might be judged. It is quite reasonable for you to say that the FiveThirtyEight forecast was a good one, and also for me to say that it was terrible, if we disagree about the kinds of data that are informative about US presidential election results. To take it to extremes again: suppose another forecaster were to have sacrificed a goat on 7 November 2016 and examined its entrails, then offered to take bets implying an

80% chance of Trump winning and 20% Clinton. Is that a good forecast or a terrible one? In this single instance, it would have dramatically outperformed FiveThirtyEight and many others.

In effect, to evaluate a single forecast we are relying on our judgement about how plausible we find the underlying model on which the forecast is based. You may have more confidence in a quantitative model that looks at national polls, then corrects for non-representative demographics, voter non-response, possible biases, etc. By contrast, an ancient Babylonian may have had more confidence in the ability of the haruspex (diviner) to interpret the goat's entrails correctly. After all, he might say, haruspicy is a respected system based on a well-developed understanding and many years of good results. Your lump of metal over there with a number on the screen is just some kind of superstitious magic.

Retrospective excuses

Retrospective excuses have the potential to completely undermine any statistical analysis of reliability. If the haruspex had said confidently, 'Trump will definitely win, unless he does something really stupid', then no doubt in the event of the alternate outcome, there would be a re-evaluation of the events leading up to the election and a retrospective pinpointing of the stupid thing that lost the election. Following the perceived failure of polling-based models after the 2016 election, there was retrospective identification of the things that invalidated those models, such as different rates of response to polls according to voting intention – the so-called 'shy Trump voter'. And the models were of course updated in advance of the 2020 election, where FiveThirtyEight offered an 89% chance of a Biden victory and only a 10% chance of Trump retaining his position.

Retrospective excuses are like the conditional predictions I talked about earlier, with the difference of not having been identified in advance. Conditional probabilities do not invalidate statistical analysis, they only describe the circumstances in which it cannot be applied. Let's say you are interested in whether the bus will be late for your regular trip on weekday mornings. First, you might make a few observations: the bus is on time on 80% of occasions and late on 20%. Following an unexpectedly bad week in which you spend a lot of time waiting for the bus in the rain, you spot a pattern and divide your data into wet days and dry days. Conditional on dry weather, the bus is on time on 90% of occasions; conditional on wet weather, it drops to 50%. You are congratulating yourself on the success of your new forecasting system when another bad week happens in dry weather. Reading the news later, you discover that there was a staff strike at the bus station all week, disrupting services. Do you include that week in the statistics for your forecast performance? How do you update your model with this new information?

One coherent answer is to turn this into another condition. The new probability of the bus being on time is 90%, conditional on dry weather and *also* conditional on no staff strike occurring (a 'get-out clause' as described earlier). This is a more informative model than the one you previously had, because it identifies a possible failure mode even if it does not assign a probability to it: we do not know how often strikes occur, only that they were (probably) not in the previous data set. Maybe we could do some more research and find out how to predict strikes, and then combine this forecast with the other one to get back to having a probabilistic forecast again.

But notice also that in order to update the model in this way, we first have to suffer the consequences of the lack of information. We could only update the forecast to take account of wet and dry days because the wet days happened. We could only update the forecast to take account of the strike because the strike happened. We will learn about other model inadequacies via model failure, retrospective excuse and then remodelling. Financial models failed to forecast the crisis in 2008 and have since been updated. Weather models failed to forecast the severity of the European heatwave of 2003 and have also been updated.

It is standard practice to update one's model constantly, with the aim of making it better by incorporating new information. For instance, we now have better weather models, which are better able to forecast the soil moisture feedbacks and other factors that made the 2003 European heatwave such an unexpectedly extreme event, with 72,000 heat-related deaths that summer. To evaluate updated models, we don't just want to wait for more data to come in. In addition, we can generate an archive of *hindcasts*: retrospective forecasts made with the current version of the model. Today's best weather model could use the data available on 1 January 1950 to make a hindcast of the weather on 2 January 1950, effectively generating the forecast that could have been made on that date if only today's model had been accessible at the time, and then comparing it with the actual weather recorded on 2 January 1950. Hindcasts are used to generate reliability statistics for new models before a sufficient quantity of out-of-sample (true forecast) data becomes available.

Yet hindcasts are not forecasts: although they do not make direct use of the data they are trying to forecast, they do make indirect use of them. The hindcast made with today's model for summer 2003 is able to forecast the heatwave well, but only because the knowledge that such an event could happen was incorporated into the development of the model. Therefore, the performance of hindcasts always gives an optimistic estimate of the performance of the model in true forecast mode. Yet the only real-world use of the model is in that true forecast mode: we do not know what might happen next year to prompt further development of the model. Will you, having made a decision today, be relieved next year to hear the retrospective excuses of those modellers that such an event could not possibly have been predicted?

This might be taken as a call for pre-emptive development of the most complex possible models to take account of all possible eventualities. It is not: there are both statistical and practical problems with attempting to do that. Instead, it is a call for humility. Probabilistic forecasts of the real world are always inadequate and we would be wise not to stake too much on their success.

Model rankings

One way to judge the performance of models is to compare them and ask which of the available models performs best in explaining the available data. Complex models, however, have numerous outputs, so to make an ordered ranking we have to find some way to collapse all of this complex output to a single number representing 'how good it is'. For example, we might decide that there are only one or two variables that really matter, then, at the points where observations are available, directly subtract the observation from the model-variable and calculate an average error. This will give a single value for each model which can then be compared with the other models. But we have jumped in with a lot of assumptions here. We might want to go back and think more deeply:

- Which variables are important? If more than one, are they equally important or should they be weighted in some way?
- Which observations are independent? A model that does well at time T may do well at time T+1 as well and therefore gain an unfair advantage by chance. If we have lots of observations in one geographical location, are we unfairly biased towards the models that happen to be good there?
- How much does being slightly wrong matter? Should there be a big penalty for getting the prediction wrong, or a small one? If we are taking an average error, are lots of small errors equivalent to one big one?
- Should we allow any consideration of physical realism? A model might perform slightly worse in prediction but have a demonstrably much better representation of the mechanisms involved: would that tempt us to rank it higher in expectation of better results tomorrow, or are we only interested in numerical performance today?
- What about the simplicity of a model? If we have two models that have equal quantitative performance on error metrics, but one of them has four free parameters and the other one has fourteen, would we prefer the simpler model on grounds of parsimony?

You might recognise these as questions we briefly tackled in chapter 2, and my conclusion here is the same: that because there are so many different ways of constructing a rank order of multiple models, we are more likely to retain maximum information by keeping a diversity of models and considering how the strengths and limitations of each can inform our thinking in different ways. To do so, we would need to write a thoughtful essay instead of simply presenting the results of various evaluation metrics. The very best scientific papers *are* thoughtful essays, not mechanical presentations of model results, although the latter are disappointingly common. Unfortunately, that process requires thinking (still a uniquely human input despite the hype about so-called artificial intelligence) and therefore does not lend itself well to automation, reproducibility or generalisation – all of which are currently in fashion in science. The cult of 'objectivity' ignores the reality of scientific methods and understanding, and encourages flattening of model information into quantitative metrics rather than informative judgements.

For instance, to answer the questions listed above would require value judgements about performance of different kinds, and these value judgements are baked into every quantitative model evaluation procedure. Usually these value judgements are not made explicitly, perhaps not even consciously. When statistical methods are taken 'off the shelf' for analysis of model results, it is easy to forget that they contain embedded assumptions that may not be the same assumptions you would make if considering it from scratch.

Of course, there are problems in the implementation of an alternative approach. Do I want every modeller to be expert in computer science, statistics and philosophy, as well as in the detail of their own field? That's clearly impossible, although it would be wonderful. Do I want to privilege expert judgement above model output? That presents the obvious risk of entrenching existing biases and inequalities of perceived expertise. But I think we are already doing this in the current use of models, so, at the very least, putting expert judgement on a more equal footing would make the biases more visible. I will have more to say about this in later chapters.

The cat that looks most like a dog

In general, models are more like other models than they are like the real world. Say we have a collection of different types and breeds of cats (models) and we want to use these to understand more about an unknown animal (reality). We can make any observations we like about the collection of cats, and we can also make a limited set of observations of the unknown animal.

Starting with the cats, we can see that almost all have four legs, fur, two eyes, a tail, upright ears, etc. The cats have a limited range of colours and patterns (none is green or blue, for example); they make certain kinds of sounds, have a range of behaviours and most of them dislike water.

Now we look at 'reality'. It also has four legs, fur, two eyes and a tail, but its ears are not upright and it is rather larger than most cats. Its colour falls within the range of the cats in the collection. It makes somewhat different noises, the behaviours are very unusual within the cat collection, and it likes water. In fact, it is a Golden Retriever. Her name is Gertie.

What do our methods of model selection do for us here? The most traditional methods will decide on a weighting of those characteristics that are most important, rank the cats from 1 to 100 and decide which is the best: they will find the cat that looks most like a dog. We then throw away all of the information from the rest of the cat collection, and use our observations about the habits and preferences of Felix, the Cat that Looks Most Like a Dog, to decide how to treat Gertie the Golden Retriever.

Other methods take an opposite approach which rules out members of the model collection based on their inconsistency with observation. Instead of singling out Felix right away, we start with our data about Gertie and initially rule out all those cats with non-golden fur. Next, we rule out the smallest cats in the collection. Now we have a set of cats (including Felix) that are all large and golden, and yes, they do look a bit like Gertie. Might we get more useful information from this set than from looking at Felix alone?

I think so. It's surely more likely that each one of that set of cats is a little bit dog-like than for Felix (the Cat that Looks Most Like a Dog)to be the most dog-like in every single respect. And we will maintain a healthy range of uncertainty, which reduces the risk of overconfidence.

In passing, it's worth noting that both Felix and the set of large golden cats *are* actually pretty informative about Gertie: she eats meat and will happily wolf down Felix's cat food; she needs water and exercise and company; a veterinary practitioner trained only in dealing with cats would be able to address many (though not all) of Gertie's potential ailments.

Having ruled out small and non-golden cats, we could next rule out the ones that have tails that are too short, and those that do not run after a stick or a ball. Like the child's game 'Guess Who?' or 'What's their Name?', each time we incorporate new information about reality, we rule out members of the collection of possibilities. Unlike in the game, however, Gertie herself is not in the line-up of cats. Very soon, if we keep making more observations and ruling out those models that are inconsistent with them, we will have eliminated *all* of the cats and will be back to having no information at all.

This is a foundational difficulty for statistical methods that seek to refine information from multiple models. These methods typically assume that we are in a 'Guess Who?' scenario: one of the available models is the correct model and our task is to identify it. When we have little data (perhaps just a fur sample), there may be many models that are all consistent with that data and we can happily use all of those models in support of our inference about reality. With a little more data, we start ruling out some models and thereby gaining greater confidence in the ones that are left, 'reducing uncertainty'. When we have lots of data, intuitively we should be in an even better position, but in practice this extra data allows us one by one to rule out *all* of our candidate models and suddenly we are back to square one. The more data we have about reality (Gertie), the more easily we can say that she is not any of the candidate models (cats).

The next tack of the statistician is to introduce a term usually known as the 'discrepancy' and then estimate or model that discrepancy itself as well, often as a random variable or random process. In theory, the model discrepancy would account for all of the difference between model and reality. But when you subtract a cat from a dog, you do not get random noise. Nor do you get random noise if you subtract real climate observations from modelled values at the same point. Both the real thing and the model are complex multidimensional systems with a great deal of structure. Their difference is also a complex multidimensional system with a great deal of structure, almost never a simple 'random error'. To assume that errors are random is to assume that our model is structurally perfect and that the only uncertainties result from inaccuracies, rather than misunderstandings. A Golden Retriever is easily distinguishable from an out-of-focus cat.

The vastness of (model) space

If it weren't already hard enough to distinguish between a handful of imperfect models with insufficient data, we also have to remember that the candidate models in our set are often only a tiny fraction of the possible models we could consider. Large and complex models may have many parameters that can be varied, and the complexity of doing so increases combinatorially. Taking a basic approach and varying each parameter with just a 'high', 'low' and 'central' value, with one parameter we need to do three model runs, with two parameters nine, with three parameters twenty-seven; by the time you have twenty parameters you would need 3,486,784,401 runs. For a model that takes one second to run, that's 776 years of computation. If you have access to a supercomputer, you can run a lot in parallel and get an answer pretty quickly, but the linear advantages of parallel processing are quickly eaten up by varying a few extra parameters.

This is a good reason for not varying π or the conservation of mass, by the way. If we're limited to choosing a small handful of the possible parameters to vary, starting with the most uncertain ones is probably the best idea. But it does mean that sensitivity analyses for anything other than the very simplest models are generally incomplete. The statistical approach to dealing with the inability to compute millions of model runs is to use what are called emulators or surrogate models, basically models-of-models that take as input the few runs you have been able to do and interpolate between them to construct a statistical estimate of what the full model would have done if you ran it with other parameter choices.

Unfortunately, the vastness of model space means that in higher dimensions, even when using an emulator to reduce the computational load, the minimum amount of real model runs needed is still prohibitive. Even if you were happy to calibrate using only the 'corners' of the space, i.e., just taking a 'high' and a 'low' option for each parameter, you still need 2^N points in N dimensions. If you don't have a model run at the very least for each of the corners, then your emulator is really extrapolating rather than interpolating into a significant volume of unexplored parameter space. In many dimensions, due to what is referred to as the 'curse of dimensionality', there are more 'corners' relative to the amount of space overall than in lower dimensions, so this actually becomes a much bigger problem than intuition suggests.

So there are real problems in assessing uncertainty in complex models, even where we use statistical methods like emulators to reduce the number of model runs we need to compute, and even if we restrict ourselves to Model Land and just conduct a sensitivity analysis over the model parameters. If there are complex behaviours somewhere in that parameter space, how plausible is it that we can even find these complexities, let alone estimate their effects? More likely they would simply go unnoticed.

Quantifying the right uncertainties?

Statistician James Berger and physicist Leonard Smith have written about the difficulties of applying formal statistical frameworks for the quantification of uncertainty in complex model outputs, raising the kinds of objections explored above. They point to the very real success of Bayesian uncertainty quantification methods for contexts such as weather forecasting and contrast it with the difficulties encountered in other situations, distinguishing between 'weather-like' and 'climate-like' applications. Following Berger and Smith, weather-like prediction tasks are ones for which:

- similar decisions, in similar contexts, are made frequently (e.g., daily or monthly);
- a sizeable forecast-outcome data archive is available; and
- the models involved have a long lifetime relative to the lead time of the forecast: out-of-sample additions to the forecast-outcome archive will soon be available.

For weather-like tasks, the statistical formalism is applicable and useful. In contrast, they define climate-like tasks as those where:

- the decision is effectively one-off, and the decision not to act, say, to gather more information, has potentially significant costs;
- the forecast-outcome archive is effectively empty (the task is one of extrapolation); and
- the model has a short lifetime compared with the lead time of the forecast.

We are back to the reference class question: weather-like models are those for which we have enough reasonable evidence to suggest that the model outputs form a reasonable reference class for the real outcome. Where did Berger and Smith go with this idea? Essentially, having made this distinction, they claim (and I agree) that for weather-like models a formal Bayesian statistical framework is a reliable way to assess and present uncertainty in forward predictions derived from models, stressing the necessity of incorporating formal descriptions of model imperfections. In climate-like situations, they demonstrate the possibility of unmodelled phenomena resulting in outcomes that are beyond the range of model outputs and therefore unquantifiable in principle as well as in practice. They emphasise the need for what they call 'uncertainty guidance': an expert opinion that clarifies whether or where the model may be inadequate, and how quantitative outputs should be used in practice. For example, for some time the guidance offered by the Bank of England probability forecasts of economic growth noted a conditionality on Greece remaining a member of the eurozone, so by implication they would be uninformative if Greece were to have exited.

What this means is that attempting to provide a full Bayesian analysis of uncertainty in a 'climate-like' situation is a waste of time if you do not also and at the same time issue guidance about the possible limitations. That could be quantitative guidance such as a definition of some probability of the actual outcome falling outside the modelled range, or it could be qualitative guidance that states some failure modes without defining probabilities. Hopefully this is sounding familiar – these are the exits from Model Land. Berger and Smith stop at that point and say that it raises further questions about how this kind of guidance should be arrived at and the role of expert judgement in constructing it. In that sense, it represents a starting point for the book you're reading now. If we want to escape from Model Land, how can we make that kind of uncertainty guidance, given our limited knowledge and socially constrained experience and thinking?

As Nassim Taleb said, probability is a qualitative subject. It contains potentially quantifiable elements relating to randomness observed in the past over different outcomes, but it also includes unquantifiable elements relating to fundamental ignorance about the system and the possibility of being mistaken in our assumptions. If we mistake the latter for the former when making a model, we will be overconfident in our predictions.

In Berger and Smith's 'climate-like' cases, we have to learn and make decisions despite not having sufficient information to do so with mathematical reliability. A mathematician or a computer might stop at that point and declare the task impossible, but let's remember that people, businesses and organisations face exactly that challenge every day and they are not paralysed; they – we – *can* make decisions despite radical and unquantifiable uncertainty. So the takeaway here is not that mathematical methods are a waste of time, although they can certainly be costly distractions in some circumstances. Rather, it is that the methods we use have to take into consideration their context and limitations. If there are uncertainties that are potentially larger than the range of variation between your models, then making a probability distribution from your models gives you quantitative information only about the next model run – the cat that looks most like a dog – not about the real world. Either way, we need to use expert judgement both to distinguish what part of Model Land we are in and to ascertain the closest exit.

5

Fiction, Prediction and Conviction

> [F]iction gives us empathy: it puts us inside the minds of other people, gives us the gift of seeing the world through their eyes. Fiction is a lie that tells us true things, over and over.
>
> Neil Gaiman, *The View from the Cheap Seats* (2016)

There is a parable about maps that is frequently retold and reinterpreted as an allegorical comment on decision-making, leadership and information. An expedition party become caught up in snow in the Alps and cannot find their way. They consider themselves lost and are resigned to death. But then one of them finds a map in their pocket; they are relieved, pitch camp and later navigate their way home. On arrival, it turns out that the map was not even a map of the Alps but a map of the Pyrenees.

What lessons might we draw from this parable? Is it that the map saved the expedition party simply by virtue of being a map, and not because it contained useful information, since it was a map of the wrong mountain range? In the management context, it is often used to support the idea that having any plan is better than having no plan: that a map that is not the 'correct' map or a model that is not the 'correct' model can still either inform or at least motivate us sufficiently well to result in positive outcomes.

But did the map actually perform its nominal function and help the unit find their bearings, despite being the wrong map, or was its physical existence simply enough to avert panic so that they waited out the snowstorm in a camp and then were able to find their own way back? In terms of drawing any conclusions relevant to real life, these two explanations would have very different implications. If the first explanation is correct, it could be an argument that our models do not have to be perfect to be useful; but in the case of the second being correct, it would be a strong argument that the utility of models is not derived from their information content, but rather from their ability to support a decisive course of action, even if that confidence is misplaced.

I will have more to say later about the implications of finding oneself in a different Model Land than intended. This chapter is about the contention that models can be useful as aids to decision-making when they are accurate, and perhaps even when they are not. The psychological aid of having a framework for thinking, and the way that models help to form narratives around different scenarios, mean that even without predictive accuracy there can still be benefits in modelling – but we do have to tread carefully.

Common ground

Discussions about the future need a sufficient basis of common ground for any speculation to make sense. If you think an asteroid is likely to hit the planet tomorrow, your basis for discussion is completely different from that of someone who expects the future to look much like today – in fact, it may be pretty much impossible for you to talk to them. Whether it is religious groups who expect an imminent ascendance of the righteous to heaven, economic or political forecasters who expect that world GDP will continue to rise at 3% per year indefinitely, or people who think that artificial intelligence will soon reach some kind of critical point and take over the world, we can all think of a group whose serious and genuinely held expectations of the future differ greatly from our own. They may be so radically and fundamentally different that a discussion with them, for example, about whether to save for retirement in 2050 or the kinds of jobs they expect their grandchildren to have, will simply have no solid common ground on which to begin.

Those are extreme examples. Yet in smaller ways all of our conversations about the future are coloured by our expectations. The ease of talking to people with whom we share assumptions, and the discomfort of talking to people with whom we do not, helps to reinforce confidence in the assumptions themselves, developing a shared narrative into something that begins to have a life of its own and can edge towards the dangerous territory of 'groupthink'.

Finding common ground on which discussion can take place between people with different perspectives is more necessary than ever in the age of the internet, increasing polarisation and global crises. Thinking about the ways in which a global discourse develops around questions like climate policy, pandemic response or financial regulation, it is clear that any basis for discussion requires some element of consensus on first principles. If you think that carbon dioxide is not a greenhouse gas and I think that it is, we will not get very far in a discussion of climate policy. If he thinks that a new disease kills 5% of those infected and she thinks the figure is only 0.1%, they are bound to disagree on an appropriate response. On the other hand, if we can mostly agree about what we think are facts (even if we are both wrong), then we can have a genuine discussion about values. We may still disagree entirely, but it will at least be clear that our disagreement results from different value judgements about the outcomes, for example, present versus future costs, or the role of government versus that of the individual, or the appropriate balance of freedom with responsibility to others.

All real-world decisions are necessarily a question of values: what is the intended outcome? Who will benefit? Who will pay? What collateral damage is acceptable? What risks are tolerable?

A particular difficulty for the construction of common ground on which to hold discussions about the future is the way in which models that project future conditions entangle facts with values. When models are presented as best-available statements of physical law, their projections are a kind of conditional fact ('if carbon dioxide emissions continue to rise, then X will happen'; 'if everyone wears a mask to reduce disease transmission, then Y will happen'). Yet they are nevertheless also dependent on other assumptions and value judgements.

In recent years, more and more of the discussion within larger-scale policy-making environments has come to be informed by quantitative numerical models that forecast detailed elements of the future. Without agreement on the basic elements of those models, opponents end up sniping at different elements of the models used by the 'other side' rather than engaging with what I think is a much more fundamental debate about values and outcomes.

Conviction narratives

In simple decision-making situations, we know roughly what outcomes will result from our actions, and the question is how to scan the possible outcomes in as efficient a way as possible to identify the actions that will result in the best outcome. There may be some uncertainty about the outcomes and this can be taken into account by including a level of risk tolerance in the definition of what is meant by the 'best' outcome. This is what is often called 'rational decision-making' and frameworks for arriving at the 'best' outcome in many different kinds of situations are well researched.

Of course, very few real-world decisions are so simple. In deciding which of two job offers to accept, for example, you are not only weighing up the quantified benefits offered by each position, but also the overall career prospects, the attractiveness of the location, the impact on your spouse or family, the effect on your existing friendships if you move away and so on. In situations of what has been called 'radical uncertainty', the outcomes are unknown, often even unknowable. It is not simply a case of doing more research in order to work out the right answer, and indeed there may be no right answer, only different outcomes. If you sit down with a calculator to weigh up all of the inputs into this decision, you will quickly find that there is no way to add a salary to an opportunity to go hiking at weekends and an extra four-hour drive to see your parents, nor does it really make sense to try to force them into this framework, except perhaps in an overall list of pros and cons.

Conviction narrative theory, developed by David Tuckett and colleagues, is a psychological theory of decision-making under radical uncertainty which begins from the assumption that human decisions are not rational in the above sense; instead, they are driven by emotion. Conviction narratives

> enable actors to draw on the information, beliefs, causal models, and rules of thumb situated in their social context to identify opportunities worth acting on, to simulate the future outcome of the actions by means of which they plan to achieve those opportunities, and to feel sufficiently convinced about the anticipated outcomes to act.

Conviction, because the process results in confidence which allows action despite uncertainty. Narrative, because of the emotional necessity of some form of explanation as justification and as a structure by which to relate the decision. What the conviction narrative allows decision-makers to do is to convince *themselves* of the reasonableness of following a certain course of action. Secondarily, it is a convenient way to explain and justify the decision to others, perhaps generating a shared conviction narrative which helps others to act in a similar way or to provide support.

The conviction narrative itself is an overall story about the future, perhaps with subplots elaborating some positive or negative details. The job in Cambridge offers a somewhat lower salary than the job in Birmingham, but the opportunities within the company are really great for someone with your skills. You'll be much further away from family, which means you'll need to spend more on childcare, but there are very good schools in the area and you could compromise on renting a smaller house for now. And there's a much better chance of your partner finding a job there as there are several companies in that industry, whereas if you take the job in Birmingham, they would likely have to commute back to where you live at the moment, or work at home – but, looking on the bright side, you'd be able to afford a nicer house in Birmingham. You've heard Cambridge is really cold, but a friend lives there and she seems quite happy, so it can't be that bad. You begin to see a picture of yourself living in Cambridge, being promoted within the new company, renting to begin with but able to buy a much nicer house in a few years' time, with your spouse also in a good job locally and your children doing well in the excellent local schools. The downsides of Birmingham start to look more important – it would be difficult for your partner to work from there – and the downsides of Cambridge find justifications: it's OK to be further from family, because you'll only need their help with childcare while the kids are little, and this is a long-term decision. A couple of days later, your friend asks whether you've decided between the job offers yet. You relate some of your thinking, although you 'haven't actually made your mind up yet' – but by the time you've finished the conversation it's quite clear to both of you which you're going to go for.

Decisions like these are not subject to the quantitative decision frameworks that guide procedures like optimisation of airline bookings, randomised controlled trials of new drugs or improving industrial processes. There is no obvious bottom-line measure of success: even if you could quantify some level of overall happiness in five or twenty years' time, it would be contingent on so many other things that it couldn't be attributed to this single decision. There is no reasonable way to assess the counterfactual: how do you know how happy you would have been if you had taken the other job? What about the effects of all those people you never met? And yet there is real ambivalence here – it is not at all obvious which of the two lives you might lead would be the better one – and a real deadline: 5pm next Friday.

If we could not construct some form of conviction narrative to settle emotionally on a course of action, this kind of everyday decision would be soul-crushingly impossible to navigate. Some people are filled with regrets about particular life decisions, true, but for the most part we learn to accept that we just have to take some decisions as best we can, without becoming bogged down in an intellectual swamp of counterfactuals and possible alternate universes. That is the 'brute force' approach taken by chess-playing artificial intelligence systems such as Deep Blue, the computer that famously beat grandmaster Garry Kasparov in the mid-1990s. It tries to look at all possible future moves, but even in a limited universe of sixty-four black and white squares and thirty-two chess pieces, all-possible-futures are so large in number that no human could possibly use this kind of strategy and even the computer is limited to looking only a few steps ahead.

Nor can humans directly use the strategy of Deep Blue's successor. AlphaZero, constructed by engineers at Google's DeepMind, searches for good strategies by playing huge numbers of games against itself and assigning estimated probabilities of winning to each move. Although in the real world we cannot possibly think through all the consequences of even the simplest actions, in a way the development of these probabilities is somewhat like the formation of a simple conviction narrative. David Silver and colleagues from DeepMind describe how the AlphaZero algorithm 'focus[es] selectively on the most promising variations', in the same way that we ignore the vast majority of possible futures and home in on a few that seem either most likely or most desirable. The only emotion that the computer can integrate, though, is a simple scalar of excitement at the prospect of winning, where human choices cannot be reduced to 'winning' either in a Darwinian sense of reproductive success, a hedonistic sense of some kind of maximum happiness or the economic sense of maximising wealth. Attempts to characterise any human decision according to these limited scalar metrics are always doomed to failure, especially if we note that they often work against each other. Having children may be the ultimate Darwinian success, but not so good for your finances; and financial success through hard work or a lottery win is by no means guaranteed to lead to happiness.

In Model Land, the computer has a clear and unambiguous aim: win the game. Outside Model Land, the game itself is unclear and to some extent we are making the rules up as we play. Living by our own rules is both an opportunity and a risk. The conviction narratives that we create to facilitate decisions also frame the way that we think about decisions, as they highlight different kinds of pay-offs. One can imagine making quite different decisions about the job search if we were aiming solely to achieve maximum financial wealth or to achieve maximum reproductive success. I don't know anyone who thinks of their own life in Darwinian terms, but research suggests that a degree in Economics can significantly weight your own value systems towards quantitative metrics and less social behaviours.

In any case, individuals can take responsibility for their own goals and values and the idiosyncrasies of their own conviction narratives. Different individuals would make different decisions in the same circumstances; some might even say this is a big part of what makes us human. Probably this variation is also what makes us so adaptable and resilient as a species: someone had to be the first to try eating puffer fish, or to discover that certain deadly plants can be eaten after complex preparation.

On the other hand, when group decisions need to be made, these differences can present a great challenge. The narrative aspect of conviction narratives helps communication within a group and to establish a shared sense of possibility, risk and expectation, assisting the social process by which a decision is reached. The conviction element, however, can be dangerous, leading to emotional attachment to decisions, entrenchment of viewpoints, and unwillingness to revisit assumptions or take into account conflicting information or views from outside the group. As Tuckett puts it, 'individuals can idealise objects of desire (for example a dotcom stock) to the point where they cannot entertain doubts about them; or their desire to feel the same as others can undermine their ability to question shared narratives'.

An engine, not a camera

Shared models of the future are what allow social discussion of imaginaries, helping to create and reinforce conviction narratives about the kinds of actions (policies, regulations) that will result in desirable outcomes, and importantly also to shape what those desirable outcomes themselves might be. Shared models include the canon of popular literature and art, from *Star Trek* to the *Ramayana*, and of course these shared models differ greatly between different communities. They provide frameworks within which to explore counterfactuals, think through the consequences of our hypothetical actions and imagine possible outcomes.

Many of these shared models are also oversimplified fictional narratives that use emotionally manipulative descriptors or even music to railroad the audience into a certain experience. That's no bad thing in and of itself: seeing the world through another's eyes is a way to develop empathy, diversify our perspectives and gain insight into how other people think. The question is whether the models (books, movies, assumptions, scientific processes) we are exposed to are sufficiently varied to achieve that, or whether they have an opposite effect of making us see only through the eyes of one group of people. As David Davies notes, 'films that present geopolitical events as clashes between forces of good and forces of evil do not furnish us with cognitively useful resources if we are to understand and negotiate the nuanced nature of geopolitical realities'.

Simplified historical and political models are manipulative in similar ways, embedding sweeping value judgements that not only reflect the prejudices of their creators, but also serve to reinforce the social consensus of those prejudices. The communities that share these models may be an entire society ('the Nazis were evil') or a subset ('the damage of Covid-19 lockdowns is worse than the disease'). Shared models bring communities closer together, but when they become reinforced into conviction narratives that exclude other perspectives, they also divide us from other communities with different shared models. There is much talk at the moment about how social media may be causing a wider polarisation in views and political extremism; in my view, a contributing factor is the way it allows us to share and reinforce our mental models and avoid positive exposure to alternative perspectives.

Other shared models (or metaphors) include the 'household budget' model of a national economy, which suggests prudent spending and saving but does not reflect the money-creating abilities of national governments. The 'exponential growth' model of epidemic disease spread in the absence of intervention was crucial in generating consensus around national lockdowns in the early days of the Covid-19 pandemic, but becomes less appropriate in each subsequent surge of virus prevalence. Perhaps one of the problems that the climate change debate suffers from is a lack of a simple shared model or metaphor beyond 'global warming' or 'global catastrophe', neither of which is conducive to taking meaningful action.

The idea that the model itself is not just a tool but an active participant in the decision-making process has been described as 'performativity'. In later chapters we will meet several examples of performative models, from the Black–Scholes model of option pricing to Integrated Assessment Models of energy and climate. Sociologist Donald MacKenzie described the Black–Scholes model as 'an engine, not a camera' for the way that it was used not just to describe prices but directly to construct them. This is a strong form of performativity, more like a self-fulfilling prophecy, where the use of the model directly shapes the real-world outcome in its own image.

There is also what MacKenzie calls counter-performativity, the idea that making a prediction of some outcome actually makes it less likely for it to happen. Counter-performative models might, for example, forecast some kind of bad outcome in a 'business-as-usual' or no-intervention scenario, with the aim of motivating change that will avoid that outcome. Examples include forecasts of the unrestricted spread of Covid-19 in the spring of 2020, which motivated lockdown and social-distancing policies that reduced the spread and therefore ensured that the worst-case outcome did not happen. Counter-performative models of this kind tend to suffer from bad press and be accused of exaggeration or overhyping the problem. Widespread predictions that the date format used by mission-critical computing systems would cause devastating failures on the Y2K rollover from 31 December 1999 to 1 January 2000 were proved entirely wrong, at least in part because of the efforts of many system administrators to ensure that important systems were unaffected or patched. Counter-performativity is evidently much harder to identify in practice, because although we can create counterfactual scenarios, we can never be sure what *would* have happened.

Performativity is very strongly in evidence in the financial policies and statements of central banks. When Mario Draghi, then-president of the European Central Bank, stated in the middle of the euro crisis in 2012 that the institution was 'ready to do whatever it takes to save the euro', that announcement in itself was enough to reassure the financial markets and avert a crisis due to lack of confidence. If a central bank were to predict a financial crisis, under any conditions we can be pretty certain that one would immediately occur. Of course, this raises questions as to what the predictions and projections of bodies with economic power actually mean. They could surely never predict a crisis, even if it did in fact seem like the most likely outcome. The forecast is a part of a narrative, and is part of the policy and intervention itself rather than being a disinterested observer. An engine, not a camera; a co-creator of truth, not a predictor of truth.

In between the two simple narratives of self-fulfilling and self-defeating prophecies, there are the other, less identifiable kinds of performativity, where things are not so simple. Instead there is a complex dance of predictions, actions, updated predictions, alternative actions and avoided counterfactuals. The model is still just as much an engine as in the simpler cases, but in different ways.

Saturn in the Ninth House

Narratives about the future do not need to be based on 'truth' in order to be useful in terms of resulting in better outcomes. An unpublished study of farmers in Mali who were given access to seasonal weather forecasts showed that they had better outcomes at the end of the farming season than farmers without access to these forecasts, even though those particular forecasts showed very little skill in predicting the actual weather. One reason for this success might be that the farmers were able to frame their decisions more effectively on the basis of risk management, or to make more confident decisions about when and what to plant. I don't want to make too much of this example, though, since there could be a selection bias here too: for example, if the most financially secure farmers were the ones who made use of the information and those who were already struggling could not manage the additional burden of considering the new input.

Thinking about other practices, ideas such as 'planting by the Moon' support gardeners by providing a clear framework for planting peas in one week and carrots in another, with the result that the peas and carrots are planted at an approximately sensible time and do not end up left on the shelf or forgotten. Often these practices are supported by semi- or pseudo-scientific justifications referring to the gravitational power of the Moon or its influence on ground water. My point is that, regardless of any justification or truth value, the framework *by itself* can be a positive influence on the actions and outcomes. Use of such a framework can convey complex information in a simple and memorable format, systematise potential actions so that they can be confidently undertaken even given uncertainty about the future, and induce us to consider (and perhaps mitigate for) risks we might not otherwise have thought of. If I had a perfect weather forecast for the spring, I would be able to decide exactly when to plant different crops given the timing of frosts, rains and warmer periods. Without such a forecast, but in the knowledge of large uncertainty, I might spend a very long time thinking about whether the last frost had passed, agonising about whether to sow tender plants now or wait another week. Having a framework that creates a conviction narrative removes this doubt – in any case, if one sowing is blasted by a late frost, it's usually possible to sow again and catch up. I have always been somewhat sceptical about fellow gardeners who make use of these systems, but after thinking it through, I am half-inclined to join them.

The use of astrology and horoscopes in the royal courts of medieval Europe is a fabulous example of a system that acted to generate and systematise insights about political relationships, manage uncertainty and support the conviction to act despite swirling layers of intrigue – and it is also a salutary reminder of the dangers of over-interpretation. Rather than acting as a definitive predictor of the future, horoscopes were used to warn about the presence of malign influences, or to decide the best time to take an action such as formalising a marriage or entering into a war. Casting a horoscope chart for an individual or for an event would reveal the positive and negative 'aspects' of the planets, interpreted by means of a complex mathematical system. Let's assume that there is no possibility of correct information being revealed in this random way. Even so, there were still potential benefits: first, it provided a systematic framework for political advice to be offered by trusted expert intermediaries, who were in general well educated, well connected and often personally very close to their lords (many also acted as physicians to the royal families). Given the arcane mathematical nature of the forecasting itself and the obvious requirement for interpretation in order to make the planetary aspects directly relevant to current events, there would be an opportunity for the advisor to bring up almost any concern, to advise caution or to impart confidence. In essence, my view is that the astrological format acted primarily as a framework within which the advisor could (more) safely discuss difficult subjects. In this way, it may indeed have been a constructive basis for positive action – but the expertise that was being made use of was in the mind of the advisor, rather than in the positions of the planets.

That's the best-case scenario. Of course, there are also many examples of much less helpful uses of astrology. As described in Monica Azzolini's fascinating book *The Duke and the Stars*, the fifteenth-century duke of Milan Ludovico Sforza became so obsessed with astrology that he began using it to determine the timing of every action, including the consummation of his marriage, which was delayed multiple times to find the most auspicious date. This invited the ridicule of his peers, who noted that he would do things that were obviously silly, such as waiting in the rain to enter a town at the right time.

Astrological predictions could be either performative or counter-performative, in the senses identified above, but they were never an indication of absolute and unavoidable destiny. In other words, where a negative prediction was given, the astrologer would ensure there was always the opportunity to avoid it by taking appropriate protective actions. It's worth noting that this also meant that no astrological prediction could be said to be 'wrong', and that these caveats were always very carefully baked in, especially where very specific predictions were made. This was important because if many people around the powerful individual also believed in astrology, either positive or negative predictions could affect the confidence of the public in that individual and thus become truly self-fulfilling prophecies. The astrologers were therefore incentivised to produce glowing charts showing the health, wealth, victory in battle and long life of their lords, because they would obviously be more favourably inclined to these predictions and extend their patronage accordingly. Serving as a further motivation for positive bias, Azzolini also describes reports of astrologers suffering retribution at the hands of their masters for making negative prognostications. A priest is said to have unwisely forecast the death of his duke and been cast into jail to starve, and a Roman astrologer who predicted the death of the emperor Domitian was himself put to death (although as you would expect in such a story, the emperor was then assassinated at the predicted time). Even if these are fabricated or exaggerated examples, the truth of the personal risk of making a negative forecast is in no doubt.

I am certainly not inclined to take up astrology in preference to mathematical modelling, and my aim here is not to imply that the two can be equated, but there are instructive similarities. One is the always conditional nature of future forecasts: if the conditions of a model are never satisfied, will we ever be able to say retrospectively whether it was 'right' or 'wrong'? Another is the potential for bias to creep in according to the funding of research: what kinds of mathematical models do we elevate to high status in different fields, and how does this reflect the priorities of funding agencies? And third is the ability of the mathematical framework to support and give credibility to the judgements of the experts who construct, drive and then interpret the models.

To return to a theme we have already met, the difficulties come when this kind of model is taken too literally, overinterpreted without the restraining hand of expert judgement or overly influenced by the social and political contexts in which it operates. Ludovico Sforza's slavish adherence to astrological timings had an overall negative influence both in practice and to his reputation. The unfortunate fates of some astrologers offered a grave incentive for others to become yes-men and public-relations operatives rather than 'impartial' or even constructive professional advisors.

If a system can support decision-making and – at least sometimes – result in positive outcomes without making quantitatively verifiable predictions, we should ask what it is about the system that is of benefit. Is it the qualitative truth offered by the minds of expert advisors, mediated through the system, or is it the framework itself that induces a constructive approach to the relevant decision questions? I am contending here that if astrology can offer the former benefit and systems like seasonal forecasting and planting by the Moon can offer the latter, we should also consider in detail the social and contextual effects of today's mathematical modelling and prediction systems, in addition to quantitative measures of predictive success. Are these systems truly an aid to positive and constructive thinking? Do they help to generate consensus about action? What kinds of actions or outcomes do they tend to promote? What kinds of incentives are implicit in the way that funding is allocated, and what kinds of expert voices are synthesised into today's models' recommendations?

Plans are useless, but planning is indispensable

I am sure many readers will bristle at the mention of astrology in a book about mathematical modelling. Let me say again that I do not equate the two as having equal predictive power on quantitatively verifiable claims. On the other hand, I hope you will agree that many real-world forecasts made by mathematical models are not fully quantitatively verifiable, because they themselves influence the future and because they are often conditional on assumptions that aren't met in practice. Having a good predictive guide to the future that also provides a useful framework for decision support is clearly the best scenario here and occasionally models do provide that. But where there is a spectrum of both predictive quality and decision relevance, it's less obvious which is the more important.

Predictions per se would not even be that helpful if they were certainties which offered no opportunity to influence any outcome. This is why the astrologers always left room for manoeuvre: the possibility of taking advantage of the most auspicious times (or failing to do so) and avoiding the most inauspicious possibilities (or failing to do so). For a decision support system to be helpful to the decision-maker, it must promote the agency of the decision-maker as well as identifying external influences, distinguish between different possible courses of action and surround this with a meaningful framework that allows the generation of a conviction narrative. It must also be *credible* to the decision-maker, and present-day astrology fails in many quarters on that count.

US President Eisenhower, who as commander of the Allied Forces oversaw the invasion of Europe in the Second World War, said that 'plans are useless, but planning is indispensable', meaning that the exact details of a plan are always liable to change in light of new information, but the process of planning, coming up with a plan and thinking about possible strategies and failure modes, is useful even if the plan itself is not followed. If we use our mathematical models to avoid having to plan ourselves and instead rely solely on their output, then we have a plan (which may be useless) but we didn't accrue the 'indispensable' benefits of the planning process. This is another reason why explainability, a concept we met earlier, is important for models. If we are confident that we can just predict-then-optimise and reap the benefits, no one needs to know how it works. But if we predict and optimise, and then something unexpected happens and the plan changes, it would be very helpful to be able to understand the planning process.

Models are much more than prediction tools

Sherwood Rowland was awarded the Nobel Prize for Chemistry in 1995 for his work on substances that deplete the ozone layer. He and Mario Molina showed that the release of chlorofluorocarbon (CFC) compounds, commonly used in refrigerants of that era due to their non-toxic nature, could result in the destruction of atmospheric ozone. Ozone is present in the Earth's stratosphere, where it absorbs some of the ultraviolet light that can cause sunburn, skin damage and cancer. Rowland and Molina made theoretical predictions about the potential for the depletion of ozone in the mid-1970s, prompting scientific and regulatory work to further assess the risks, find alternatives and organise political interventions. In the mid-1980s, their prediction was confirmed by observation (despite modelling studies in the interim suggesting that the effect would be relatively small), and the 1985 Vienna Convention on the Protection of the Ozone Layer demonstrated global political appetite for regulation. The story from that point forward is well known: the risk of thinning the ozone layer and exposing whole communities to greater ultraviolet light became public knowledge – children of the 1980s and 1990s will remember learning about it in school; further scientific work resolved doubt about the chemical processes involved; technological developments made alternative, less ozone-depleting chemicals available; and the Montreal Protocol, an international agreement to limit the use of CFCs, was signed in 1987. The United Nations even celebrates an International Day for the Preservation of the Ozone Layer each year on 16 September.

The Montreal Protocol is widely recognised as a remarkably effective and relatively uncontroversial translation of scientific discovery into political action, and although there is still some illicit production of CFCs, the problem is certainly a fraction of what it could have been in the absence of regulation. With the decline in the emission of long-lived ozone-depleting substances, the thinning of the ozone layer has slowed. Optimistic projections suggest that it could even recover to 1980 levels by the middle of this century; less optimistic projections point to growth in short-lived ozone-depleting chemicals and the increasing role of nitrous oxide, which has a similar effect.

As Rowland said: 'What's the use of having developed a science well enough to make predictions if, in the end, all we are willing to do is stand around and wait for them to come true?' If models are nothing but prediction engines, then they are only useful for those who benefit from making a correct prediction, such as financial traders working in something like a pure information market. But for those of us who live in the real world rather than in Model Land, the benefit of our model is the opportunity it gives us to choose actively between different futures.

The Montreal Protocol was successful not just because of the accuracy of the models involved. In addition to the confidence that CFC emissions would cause ozone depletion, there was clear agreement about the consequences for human health and that the benefits of action in terms of avoiding skin cancers and other potential damages would outweigh the costs. If the technological costs of replacing CFCs had been greater, that consensus could easily have been weaker. The acceptability of the model was a function of the relatively simple context and the low costs of action: this is, of course, in direct contrast with the situation for climate change. The model, although a complex chemical analysis, is essentially one-dimensional. The more CFCs are emitted to the atmosphere, the greater the damage to the ozone layer and the greater the potential for harm. Prediction of the exact amount, beyond a significance level of 'yes, this will have real effects', was not particularly important. In part, this was because of the consensus for action: if there had been no action, then the question of predicting the subsequent impacts would have become much more important.

The best way to predict the future is to create it

If we take models to be guides and aids to thinking, rather than simple prediction engines, they have a much wider sphere of potential use and applicability. That has been my overarching message in this chapter. Models are not simple tools that we can take up, use and put down again. The process of generating a model changes the way that we think about a situation, encourages rationalisation and storytelling, strengthens some concepts and weakens others. The question of whether a model generates accurate predictions is important where it can be evaluated, but it can in certain circumstances be secondary to the way that it is used for decision support.

In this view, models are frameworks on which to hang stories and emotionally laden narratives about the future. Models facilitate conversations between stakeholders about the potential for different kinds of action, including helping to navigate boundaries between people with different levels of power or information. More than predicting the future, they help to create the future as active participants in a social thinking process. Models have power.

6

The Accountability Gap

> And oftentimes, to win us to our harm,
> The instruments of darkness tell us truths,
> Win us with honest trifles, to betray's
> In deepest consequence.
>
> William Shakespeare, *Macbeth* (1606)

The aim of creating any model is to understand the nature of reality better and thus be able to make better decisions in the real world that you inhabit. Even if you are a theologian or a string theorist (perhaps they are not so very different), that still holds true. But if the aim of models is to inform better decisions, then there is an unavoidable question of defining what we mean by a better decision, and this is not trivial even for seemingly quite trivial questions. The well-known 'trolley problem' is one philosophical attempt to grapple with this problem: there are three people tied to one branch of a railway line and one person tied to another branch, in such a way that the passage of a train would lead to their certain deaths. A train is coming and you are at the points, which are set so that the train will kill the three people. You have the choice either to leave the points as they are, or move the points and kill the single person instead. There are no other options. Beyond the crude utilitarian question of numbers, the simple addition of an element of agency and responsibility has given this question long-term staying power in philosophical discussion and more recently as a popular format for internet memes shared on social-media platforms. When we are talking about decision-making, a utilitarian calculation is rarely the only consideration. More complex decision questions involve responsibility and moral judgement as well as, for example, emotion, politics, social power, relationships or aesthetics. These are difficult for models to encompass but are highly significant considerations.

When I say that models inform decisions, I do not mean that models do or should *make* decisions. Humans make decisions, and information gained from models about 'the consequences of action A relative to action B' is only one of many inputs to any decision-making process. Although, as I will discuss, many models are put in positions of unwarranted power when humans abdicate their decision-making responsibility.

Models and hidden value judgements

As we have seen already, the more obvious value judgements encoded in models are those of priority: what do we decide is important enough to put in a model, and what is unimportant enough to leave out? Some of those assumptions are scientific in nature: the choice not to represent the colour of a cannonball does not affect the projection of its terminal velocity when it is dropped from the Leaning Tower of Pisa and is a sensible choice based on both theory and observation. Others are clearly socially determined: the choice to put a lot of time and effort into modelling the spread of the SARS-CoV-2 virus rather than the spread of cat pictures on social media reflects the subjects' relative importance to society at large and is also pretty uncontroversial.

Other choices are pragmatic, and it is these pragmatic ones that we should perhaps examine most carefully because they can reflect value judgements and experiences that are shared by the modellers but may be unrepresentative of wider society. If we find it computationally tractable to model the atmosphere but not to model the carbon cycle, is it reasonable to leave that out on pragmatic grounds? If we find it easier to model virus-susceptible populations as a collective of identical agents rather than disaggregating by sex, racial background or social class, is it reasonable to make the pragmatic assumption that this will not matter? People who make models are primarily well-educated, middle-class individuals, often trained in a certain way that values what they perceive as scientific detachment and therefore seeks to suppress value judgements and make them less visible. Their choices reflect the social norms of the modelling environment.

The target of the trolley problem memes mentioned above is the incommensurability of value judgements. What if the three people on one rail are terrible criminals and the singleton on the other is a highly respected and productive member of society? What if we are in fact trading off biodiversity for economic gain, air quality outside a school for lower commuting times or the quality of a personal relationship for higher productivity at work? As we saw in previous chapters, some modellers would like to be able to put a single numerical value on each effect and treat them as commensurable quantities that can be meaningfully added to or subtracted from each other. In this framework, a loss of air quality outside a school leading to a certain number of cases of respiratory disease or asthma could be fully or more than fully compensated for by the financial benefits of reduced commuting times. Typically, this commensurability is achieved by financialisation: putting a dollar value on everything. This is the underpinning logic of cost-benefit analysis, of 'ecosystem services' and of carbon offsetting. In some cases we are even trading off things that do indeed have the same units, but different distributional qualities: carbon emitted by one person, burning jet fuel in the engine of a frivolous flight to Las Vegas, compensated for by carbon sequestered in a bit of the Amazon rainforest that a small-scale farmer agrees not to cut down. The logic of offsetting makes sense in a globalised zero-sum economic framework that treats every dollar as equal, every carbon atom as fungible and every attribute as infinitely substitutable. Sometimes, where a dollar value is just too crude, we find alternative commensurable units such as Quality-Adjusted Life Years which perform the same operation of reducing trade-offs to quantitative comparison.

Mathematical models don't need to do this: we can always choose to keep incommensurables separate. That the contrived trolley problem is discussed at all is a bizarre and even somewhat morbid symptom of an obsession with quantifying, comparing and judging. But, again, the social norms of the modelling environment do prioritise comprehensiveness, generalisability and universality. All of these speak in favour of slicing, dicing and weighting the multiple outputs of a model, or of many models, in order to be able to present them on the same chart. Cognitive scientist Abeba Birhane and colleagues looked at one hundred highly cited papers in machine learning, finding that 'papers most frequently justify and assess themselves based on performance, generalization, efficiency, researcher understanding, novelty, and building on previous work', but that 'societal needs are typically very loosely connected to the choice of project, if mentioned at all, and that consideration of negative consequences is extremely rare'. The same justificatory norms can be found across science, though in many ways the machine-learning community is particularly extreme.

In pursuit of comprehensiveness, models also have a tendency to become more complex over time. While a model of a shot-put throw may reach a limit at which further development makes no discernible difference to the output, models of complex systems like the economy or the climate can be continually extended to represent finer detail or different assumptions about behaviour in different circumstances. Due to the Butterfly Effect, and the similar Hawkmoth Effect which I will describe later, even infinitesimal 'improvements' to the model can continue to have significant influence on the predictions it makes, and improvements to the physical realism of a model do not necessarily go hand in hand with improvements to its empirical accuracy.

The drive to complexity introduces problems of its own. At some point, a model can become so large that no single person would actually understand all of its systems, or be able to enumerate the judgements underlying the choice to set certain parameters within given ranges. Where small models can be tuned by humans tweaking parameter values, larger models are only tuneable using automated approaches that define upfront the goal of optimisation. In many ways this is a good thing; there is certainly a lot to be said for writing down the aim explicitly. That's Step 1. The harder part, Step

2, is critiquing that aim with respect to the value judgements, socio-political standpoints or the attitudes to risk that it implies.

Reliability is not utility

Defining an aim to which a model can be tuned is equivalent to defining its purpose and its value. What is the value of a model? If you listen to scientists who make models, you'd be forgiven for thinking that the value of a model is solely a function of its ability to make reliable predictions. In that case, you could even put a financial value on it. First, you would need to say what the model helps you decide to do (Plan A). Then, you need to say what you would have done without the information provided by the model (Plan B). Finally, you need to define a scale that values how much better the outcomes of Plan A were than those of Plan B would have been. Let's pretend we could do that. Say I am an apple grower and a high wind is forecast which would tear down and damage the fruit (Plan B), but having the weather forecast I am able to choose to harvest a few days earlier (Plan A) and sell my apples at a higher price (20p per fruit) than they would have fetched if damaged (10p per fruit). This is really the value of the single forecast, since the model might make lots of forecasts that would all have some value, so then the value of the model itself to the grower would be a sum over all the times it could be used.

But my little sketch of the apple grower is in Model Land; in the real world, there are lots of other contributing factors. We could incorporate into this analysis the imperfect reliability of the weather forecast, the fact that everyone else has also seen the forecast and wants to hire the harvesting machines at the same time, the contingency that some fruit-pickers aren't available this weekend because they are attending a wedding, and so on. That done, the value of the forecast is changed (perhaps in either direction) relative to the sketch, because it interacts with so many other factors. We have assumed that maximising profit is the only bottom line, but if a change of plan would impact relationships with customers or employees, that might need to be taken into consideration as well.

If that is a more realistic conception of the input of science to a decision, then the value of the model is not so much in its absolute reliability as in its integration with the decision-making context. Maybe like the option traders we will meet in the next chapter, the use of the model is a benchmark rather than an ideal, a mental starting point onto which the complications of real-world existence can be built even if they negate the internal assumptions of the model itself. Then the absolute numerical reliability of the model is not in itself necessarily critical to its practical use as decision support.

On the other hand, those who like to quantify the value of a model are typically those who want to sell it to you.

Experts and expertise

Who makes models? Most of the time, experts make models. Hopefully, they are experts with genuine expertise in the relevant domain: a volcanologist making a model of Mount Pinatubo; a marketer making a model of how people respond to different kinds of advertisements; a paediatrician making a model of drug interactions in child patients.

When experts create models, they are using their expertise to make judgements about relative importance and causal relations, which are then encoded into a structure that can be independently explored by someone else. The expert has essentially complete power over the model and decides everything that goes into it. Their chosen structure may enforce some forms of logical consistency such as arithmetic relations (1+1=2) or the conservation of mass. Setting out to create the best possible model, the very first attempt of the expert will be what I will call a 'first-guess model', which reflects only the information they have gathered from outside Model Land.

The first-guess model is very rarely the stopping point. Although it is a direct product of their expertise, it might in some ways think quite differently from the expert. It might have unintended errors or defects – bugs – and it might also turn out, on playing with the model, that some of the intended idealisations are actually too crude or could be remedied with further improvement. Maybe it doesn't quite do what the expert wants it to, and they need to change things and see what happens as a result. Maybe they use the model to learn about how different parts of the system interact with each other, to test alternative hypotheses and explore counterfactuals.

This is a completely normal and legitimate stage of model development. But the difference is that at this point the expert is using information from within Model Land to inform their understanding of reality. The expert shaped the model, and now the model is beginning to shape the expert. Interacting with the model starts to influence how the expert thinks about the real system – we will look at some more detailed examples of this later. So, instead of drawing separate boundaries around the model and expert, we need to consider model and expert as a single system. Beyond that, when the model created by one expert begins to inform the judgement of other experts, perhaps we should draw a wider boundary around a thinking system – what some call a 'cognitive assemblage' – of experts and models together. If we have a flow of information from expert to model and a second flow from model to expert, then we have a feedback loop.

How might this feedback loop be troublesome? Well, it depends on how we interpret the outputs of the models. If we are happy to think that the model outputs are as subjective as the expert judgement on which they are based (sometimes very speculative, sometimes very directly tied to observation), there is no conflict. If we would like to believe that the model is in some way accessing or estimating an objective truth separate from the expert judgement of its creator, then the existence of this feedback loop undermines the logic of inference. First, it undermines the semi-formal use of expert judgement to evaluate model output. By this I mean qualitative model evaluations determining that one pattern of behaviour is 'more realistic' than another, or that outputs lying outside a particular range are 'unlikely'. For example, if all of our previous models have shown values of a certain parameter between 1.5 and 4, then we will naturally examine more closely (and perhaps tweak or recalibrate) those that show it to be outside that range. This generates an effective bias towards the 'accepted' values. Second, it undermines quantitative statistical methods for model analysis. If nine out of ten models do a particular thing, does that mean they are 90% certain to be correct? To make an inference like that, when models are so interconnected, would be, as Ludwig Wittgenstein put it, 'as if someone were to buy several copies of the morning newspaper to assure himself that what it said was true'. A more useful comparison would be to take it as though nine out of ten experts had agreed on that thing.

At which point we can make a very rapid exit from Model Land: instead of doing some fancy statistics about nine-out-of-ten models, we can return to a more accessible level of common intuition. Why might nine out of ten experts agree on something? It may indeed be because they are all independently accessing truth, and this will give us more confidence in the outcome. Or it might be that they are all paid by the same funder, who has exerted some kind of (nefarious or incidental) influence over the results. Or it may be that they all come from the same kind of background and training so that they approach the question in a certain way and don't think of other perspectives. Depending on your political orientation, your own scientific level and your trust in the workings of expert knowledge, you may reasonably come to different conclusions. If you ask experts to give an opinion on whether 'the vaccine works', 'the climate is changing', 'tobacco does not cause cancer', 'the world will not end next year', 'social media are harming children's wellbeing' or 'a foreign power interfered with that election result', what you choose to believe is transparently going to depend on who you take to be an expert. For me, nine out of ten experts may say one thing. For you, nine out of ten experts may say the opposite.

There is no contradiction here. There would be a contradiction if we believed that expertise were an inherent property that is either attained or not by each potential commentator, perhaps through some kind of exam or qualification. But it isn't. Expertise is in the eye of the beholder. John may think it trivially obvious that the findings of a medical expert on vaccine efficacy are trustworthy. Jane may think it trivially obvious that the findings of a medical expert are compromised by their financial interests. John may listen to experts who say that action on climate change is an unnecessary cost that threatens the economy. Jane may listen to experts who say that climate change is an unnecessary risk that threatens the economy. We need not descend completely into anarchic relativism here: clearly there are some opinions that are actually contradictory to observation and therefore wrong. But I think we all need to take seriously the possibility that other people have opposing views because they have thought carefully about it and because they have different interests and values, rather than because they are wrong or evil. If someone says that climate change is not happening or that Covid-19 does not exist, they are contradicting observation. If they say that action to prevent climate change or stop the spread of disease is not warranted, they are only contradicting my value judgements.

As such, most of these are social disagreements, not scientific disagreements, although they may be couched in the language of science and framed (incorrectly) as a dispute about Truth. Whatever the case, the concern is not so much about the scientific result as it is about the kinds of action that might be precipitated by accepting one side or the other. These questions cannot be resolved simply by looking at data. In order to defuse the disagreement, other actions are necessary: referring back to our list of reasons why experts might agree, we need to construct a system that promotes more widespread feeling that experts are trustworthy, including addressing the possibility of conflict of interest directly and ensuring that experts do not all come from the same political and social fold. So, for instance, a wider variety of voices is needed in both medical and economic arenas to reduce concerns from specific groups of people that they are being shepherded in the name of science towards a future they do not particularly like, have not had a hand in creating and from which they will not benefit (while others do). Falling back on the justification of authority is simply insufficient here and demonstrably does not work. The authority of expertise only exists when there is trust in expertise, and that is again a social question, not a mathematical one.

Let's skip back in the general direction of Model Land. If models represent the opinions of experts, how can models be more trustworthy? Well, they need to do the same sorts of things, for the same reasons: be transparent about value judgements; declare conflicts of interest; be owned by a wider variety of experts. If we are serious about addressing lack of confidence in science, it is necessary for those who currently make their living from and have built their reputation on their models to stop trying to push their version of reality on others, specifically encourage modelling efforts (however trivial) from underrepresented groups – in particular those with different political views about desirable outcomes – and acknowledge that decision-making requires value judgements as well as predicted outcomes. And yes, that's a big ask.

Have people had enough of experts?

As mentioned before, it is a fact of life that British people like to talk about the weather. On 15 October 1987, BBC weather forecaster Michael Fish presented a TV bulletin. He said:

> Earlier on today, apparently, a woman rang the BBC and said she heard there was a hurricane on the way. Well, if you're watching, don't worry, there isn't – but having said that, actually, the weather will become very windy. But most of the strong winds, incidentally, will be down over Spain and across into France.

Hours later, hurricane-force winds were recorded over the South of England, twenty-two people lost their lives and the 'Great Storm' caused upwards of £2 billion worth of insured damage. Was Michael Fish wrong? Technically it wasn't a hurricane but an extratropical storm, though that was probably of no comfort to those whose property was destroyed. One forecast like this, in the minds of the viewers, more than outweighed any modest continued success in predicting tomorrow's rain, and Michael Fish has been infamous for that one forecast ever since. In the aftermath of the storm, the failure of the models was also reviewed and found to be partly due to lack of sufficient observational data from the North Atlantic. We now have a much more robust observing system (and better models) and thus generally a much better chance of predicting this kind of event.

That is the context in which Andy Haldane, former chief economist of the Bank of England, called the institution's failure to predict the 2008 global financial crisis a 'Michael Fish moment'. During a visit to the London School of Economics in late 2008, Queen Elizabeth asked, 'Why did nobody see it coming?' With hindsight, there are plenty of answers for Her Majesty. Many people were lulled into a sense of security by the 'Great Moderation', a long period of relatively stable growth and rapid recovery from short-term downturns. Those who did see something coming, like Cassandra were not believed. Even if they had been, the likelihood is that the solutions required would have been systemic and long-term rather than immediate heroic action, which in such conditions may prompt the crisis rather than avert it. But again, these explanations offer little comfort to those who lost their house, their job or their pension in the 2008 crisis, or who have struggled in the protracted period of economic uncertainty that has followed.

In the run-up to the referendum on whether Britain should leave the European Union, British politician Michael Gove said in a TV interview, 'I think the people of this country have had enough of experts from organisations with acronyms, saying that they know what is best and getting it consistently wrong.' It's unverifiable whether the experts have got things 'consistently wrong' – who's to say whether things might not have been even worse if another line had been taken? But I think the issue here is not the rightness or wrongness of any forecasts; rather, it's the perception that the interests of experts are not the same as the interests of the ordinary people. Weather forecasters do not lose their jobs after a poor forecast; in fact, the Met Office got more investment to improve its methods. High-flying bankers who failed to prevent the financial crisis have continued their careers, analysing the reasons for the failure and managing the recovery – with the notable exception of Icelandic bankers, some of whom were prosecuted and even jailed for crimes including fraud and insider trading. As for Brexit, neither the dire catastrophisations nor the utopian dreams have yet been realised. It was humorously noted by science-fiction writer Arthur C. Clarke that 'for every expert there is an equal and opposite expert', and it is unfortunate that 'experts' do tend to be trotted out on both sides of any contemporary debate, with little to distinguish one from another even where they have completely different levels of qualification.

If some of the British people have 'had enough of experts', it's certainly not only a British phenomenon: we can see the same suspicions at work in the US, in particular in the words and actions of former President Donald Trump. His we-have-our-own-experts approach, as a 2020 feature in the prestigious science journal *Nature* put it, 'devalues public trust in the importance of truth and evidence, which underpin science as well as democracy'.

Models, if they are an expression of certain kinds of expert judgement, have the same problems. How is the non-expert supposed to be able to distinguish between the pronouncements of one model and those of another 'equal and opposite' model which says something completely different? In the debates about Covid infection numbers given different possible actions, I think we can see a narrative emerging that people have 'had enough of models' too.

The accountability gap

One good thing that comes naturally from the above view of models as being personalised judgements belonging to experts is an inherent accountability. Who is responsible for the predictions of a model? The person who made it, of course. But that isn't usually the way it works. In practice, models seem to be attributed a kind of semi-autonomous agency of their own which is separate and distinct from the judgement of the modeller. In the 1980s, Michael Fish took the flak for his wrong forecast. If a weather forecast is wrong now, it's interpreted as the fault of the model, not the person who happens to read it out on TV.

So there is a gap here between expert and model. It is a way in which models, as Paul Pfleiderer put it, behave like chameleons behind which accountability is lost. His argument was that experts can pretend their models are policy-relevant when asking for funding and support, but disclaim responsibility – 'it's only a model' – when their recommendations turn out to be suboptimal. I would take this a bit further. It's no use for experts to make recommendations in Model Land, because that's not where the rest of us live. If I run a model that says that the temperature tomorrow will be 28°C plus or minus 3°C with no rain, and give you a confidence interval taken directly from the model, then I am still in Model Land but pretending to be relevant to the real world. When it turns out to be 16 °C and raining tomorrow and your plans for the day are ruined, I cannot then make the chameleon move and say, 'It was just a model and I always knew it could be wrong.' I cannot take the credit for success and also deny responsibility for failure. Of course, sometimes even my best guess will be wrong – but that's on me, not on the model, and my attitude towards confidence intervals should be informed by how mad you will be if I get it wrong.

There is more difficulty when it comes to large models and systems of models that are the work of more than one expert, since they cannot be taken to share all opinions monolithically. The value judgements that underpin a complex financial model or a numerical model of global climate are not the work of a single person – as such, those judgements may not actually 'belong' to anyone and may not even be internally consistent between different parts of the model. This is a second accountability gap, and again it leads to the model being attributed its own kind of semi-autonomous agency in the absence of a single owner.

In model-informed decision-making there can often be this kind of accountability gap. Where does the buck stop? When asked why a certain decision is made, the chain of responsibility may lead back to a mathematical model that said this outcome would probably be better than another outcome. 'We were following the science' is a common refrain: in the best case this results in high-impact decisions with no owner at all, and in the worst case any failures will be weaponised against science and the scientific method itself. If economic models fail to encompass even the possibility of a financial crisis, is nobody responsible for it? Who will put their name to modelled projections?

In my view, institutions such as the IPCC should be able to bridge this accountability gap by offering an expert bird's-eye perspective from outside Model Land. The authors of the IPCC's reports are named, and they are respected as 'experts' by a reasonably wide community. In some sections they merely report model results, but in key areas they do offer additional expert judgements, arrived at by consensus, about the degree to which model results are judged to be reliable.

The social legitimacy of science requires that such assessments are made, publicly, to the best of our ability. If they are to be widely credible, they also need to involve a variety of voices – it matters who those people are.

Humanitarian action in disaster-stricken areas is another case in point. The traditional responsive mode of humanitarian action has been to wait until some disaster occurs – say, a hurricane – then send in aid to the people affected to help them deal with the aftermath. As mathematical models have become better, there is a genuine opportunity now to see some hazards approaching. The track and intensity of a hurricane can in some cases be forecast several days in advance with reasonable accuracy, giving an increasing window of opportunity to act in anticipation of a crisis by evacuating those at risk, strengthening houses, securing emergency supplies or doing whatever is most needed at the time. The same goes for other potentially predictable hazards such as heatwave, flooding, conflict associated with elections or surges in infectious diseases. But if a mathematical model is used to predict the event and respond pre-emptively, there is always some uncertainty about the future and therefore there is always a chance of taking the wrong action. A hurricane might swerve unexpectedly to hit a different island or intensify at the last minute to be stronger than forecast. The possibility of taking either too little or too much action due to an incorrect forecast introduces a new variable to the decision-making. Taking too little action is a humanitarian failure; taking too much is a waste of resources that could have been used elsewhere. The prevailing rationale for forecast-based action is that overall it can reduce the costs and impacts of potential hazards despite this new uncertainty, but this is by no means guaranteed.

There is even a move towards automating decisions to release large amounts of money based on forecasts of such events. In the interests of maximising the window of opportunity, the argument goes, it is necessary to have predefined triggers that can be activated quickly and without ambiguity. But it is a lot to expect a very wide community of practice to define all of this in advance, on the basis of limited and imperfect models and a scarcity of data with which to calibrate them, and to generate sufficient confidence in the overall benefit of the system to continue to stick by it even when the 'wrong' decisions are being made regularly. Not least because each of those 'wrong' decisions means loss, damage, injury and even death that might (in another universe with a better model) have been avoided. In practice, especially when the technical skills involved in constructing such a system are limited to a small number of people, the designers have a balancing act to perform in order to make the most of the information that is available, while maintaining the credibility and legitimacy of model-based methods against these challenges.

The human that looks most like a computer

As we are talking about decision-making, I want again to distinguish models from algorithms, such as those described in Cathy O'Neil's great 2016 book *Weapons of Math Destruction*. Algorithms make decisions and as such they directly imply value judgements. Where I want to continue that argument is in noting that mathematical models themselves, which are used to project if-we-do-X-then-Y-will-happen, contain value judgements *in addition to* the value judgements that algorithms make about the different possible outcomes (Y).

Let's take the example of the solvency approach to risk management in insurance, which states that insurers must hold capital for the worst event likely to happen in 200 years. An algorithm for deciding what level of capital to set is based around that 1-in-200-year event, which implies a risk attitude and value judgements about the appropriate balance between the risk of not being able to fulfil all contracts versus the profitability of insurance businesses. In addition, the model that might be used to *determine* that 1-in-200-year event size also contains assumptions that amount to value judgements about the acceptable risk of being wrong, and these judgements are made outside Model Land, by the experts who create the model.

It's becoming more and more common across the board for decisions to be outsourced to models. Doing so assumes that the needs of the decision-maker can be adequately represented by a model and that those priorities won't change over time in a way that can't be represented. The difficulty arises when, for example, the model fails to trigger humanitarian action for an event that everyone can see is going to be catastrophic: then the reputational damage of failing to act may be more weighty than the commitment to follow the model.

More generally, we have the problem of autonomous systems, which some call 'artificial intelligence' or AI. It seems to me that rather than AI developing towards the level of human intelligence, we are instead in danger of human intelligence descending to the level of AI by concreting inflexible decision criteria into institutional structures, leaving no room for the human strengths of empathy, compassion, a sense of fairness and so on. Is the 'best' decision always the one that maximises financial return and the 'best' human the one who looks most like a computer? The concept of 'rationality' in economics would have you believe so. What would happen if you brought up a child to believe that maximising a single variable was the point of life? Tongue in cheek, one might say you would be raising an economist (luckily, this stereotype is unfair and most economists are humanly irrational like the rest of us). More seriously, you would be raising a psychopath. Much of the world's literature, both secular and religious, speaks about the importance of non-rational and non-financial obligations and relationships that make our social world. How should we quantify the principle of 'love thy neighbour' in order to instil it into an artificially intelligent autonomous system? Is it then a competition in which the winner is the one who loves their neighbour the most?

AI is fragile: it can work wonderfully in Model Land but, by definition, it does not have a relationship with the real world other than one mediated by the models that we endow it with. Decision-makers sometimes mimic AI, as when they insist on an algorithmic maximisation of expected financial return. Decision-makers sometimes pass the buck to AI by wholly outsourcing decisions to algorithms based on models. When decision-makers do these things, however, they make themselves fragile in the same way that AI itself is. They have blind spots in the gap between Model Land and real world: both unavoidable mathematical blind spots such as the inability to extrapolate safely beyond known conditions, and social blind spots such as the inability to consider or value non-quantified outcomes.

The algorithm that chooses whether to show you an advertisement for a fancy holiday or a new bicycle is based on a model that draws a boundary around you as a consumer, the medium and the company that is advertising to you, aiming to maximise profit for the medium carrying the advert by generating traffic to the advertiser's own site. This model does not consider factors such as your mental health, your ability to pay for the item in question or the ecological impact of the purchase. In short, it does not *care* for you, for your relationships or for your future. Doing that within a model is not in principle impossible, it would just introduce too many variables, take too long to compute and ultimately reduce profitability – at which point it is in conflict with the business model that states that financial return should be maximised. There is a classic parable illustrating the dangers of AI, which offers a scenario in which someone creates an intelligent paperclip machine with the aim of producing as many paperclips as possible, as efficiently as possible. The machine performs incredibly well, producing vast numbers of paperclips from the raw materials at hand; it then co-opts other resources into the production of paperclips and eventually takes over the world, subsuming everything and everyone into the pursuit of more and more paperclips at the expense of everything else on Earth. Clearly ridiculous, right? This should be the point where a Hollywood hero turns up, switches the machine off and saves the day.

And yet we already have the equivalent of a paperclip machine doing exactly this. It's not maximising some physical item – it's maximising numbers in electronic bank accounts of corporations and the wealthiest individuals across the planet. Even the billionaires who make a decent effort to give away their money end up accumulating more and more. Corporations are not even able to attempt to give away their money because they are bound by the so-called 'fiduciary responsibility' to return cash to shareholders at the expense of everything else. One day, I hope, this fiduciary responsibility will include a commitment to the children and grandchildren of those shareholders, to the ecosystems that support us, to the fundamental morality that makes us human. In the meantime, our capitalist paperclip machine is destroying our life-support systems at least as effectively as if it were producing paperclips in the manner described above. It generates plastic waste, carbon pollution, denuded soils, hollowed communities, divided societies – while the number of zeroes on the ends of those big bank accounts just goes up and up. And this AI has even anticipated the coming of the Hollywood hero and is doing its best to sabotage her efforts whenever she tries to resist.

What happens if we make capping global average temperature the sole goal of climate policy? As I've mentioned, climate policy that does not consider the distributional and moral impacts of climate change and climate policy together runs the risk of continuing to privilege the systems that have got us into this mess. An artificial intelligence, or a human with similarly limited motivation, would no doubt look at the system and conclude that the most efficient solutions are large-scale technical and coercive possibilities like geoengineering, population control, virtual reality environments and rationing by price. The models for climate policy which assume that individuals are financial maximisers, and cannot be expected to do anything for others or for the future that is not in their own narrow short-term self-interest, are self-fulfilling prophecies. They limit the kinds of decisions we are even able to consider as possibilities, let alone model in detail. A human who does not look like a computer, and is not socially pressured to pretend to be a computer or to outsource decisions to a computer, can meaningfully integrate concepts such as care, love, responsibility, stewardship, community and so on into decision-making. Maybe a mental model which assumes that individuals also have values and can be trusted with the responsibility to care for the future in the way that we honour our ancestors for having done so can also be a self-fulfilling prophecy. By that I don't mean to say that I think there are easy solutions. Artificial intelligences and autonomous systems are 'easy' solutions that avoid the hard work of defining inclusively what it is that we care about and how we want society to look.

Mathematical modelling is a hobby pursued most enthusiastically by the Western, Educated, Industrialised, Rich, Democratic nations: WEIRD for short. Given that computers have been designed and implemented by WEIRD people, it's no surprise that computers look more like some humans than others. Abeba Birhane writes that 'ubiquitous Artificial Intelligence (AI) and Machine Learning systems are close descendants of the Cartesian and Newtonian worldview in so far as they are tools that fundamentally sort, categorize, and classify the world, and forecast the future'.

The kinds of modelling methods that are most used are also those that are easiest to find funding for, and those that are easiest to get published in a prestigious journal. In this way, formal and informal scientific gatekeeping enforces WEIRD values onto anyone who wants to do science: those who make it to the top have tended to be those who do things in the accepted way. The power gradients seen in model-based sciences thus reflect the power gradients in the rest of our society. Jayati Ghosh describes this in economics:

> Much of the mainstream discipline has been in the service of power, effectively the power of the wealthy, at national and international levels. By 'assuming away' critical concerns, theoretical results and problematic empirical analyses effectively reinforce existing power structures and imbalances . . . Economic models that do not challenge existing power structures are promoted and valorised by gatekeepers in the senior ranks of the profession.

It wouldn't be so much of a problem if these were just siloed academics arguing in their ivory towers about the best abstract way to represent the universe. As I hope this book makes clear, though, we are all affected by the way mathematical modelling is done, by the way it informs decision-making and the way it shapes daily public conversations about the world around us.

Don't throw the baby out with the bathwater

If there is a problem in trusting models too much, there is equally a problem in trusting models too little. Where they do contain useful information and provide a helpful framework for thinking about the future, we don't want to throw the baby out with the bathwater. Should we conclude that certain kinds of mathematical models are irredeemably contaminated by a certain value system which makes them only nefarious agents of elite control, and therefore throw them out, never to look at them again? I think that would be an overreaction, if an understandable one. First, we need to think about what we would replace them with. Other models, with different perspectives, will be equally value-laden. We do not want to reproduce the hubris of the current system by deciding a priori that a different set of values is the only right one, so while replacing the current set wholesale might be cathartic, it might also be unhelpful.

Second, the amount of effort put into developing some of these models is phenomenal. Imagine a tree of possible model structures, where each branching point represents a different choice made about how to abstract observations into a model. At present, in many fields, one single branch is huge and outweighs all of the others, with an array of tiny branches at the very tip but no major forking further down. Would this tree of knowledge thrive and fruit more effectively if we cultivate the other branches, or if we cut it down entirely and hope that it will sprout more evenly from the roots? I think there is a risk that the tree may not survive a hard pruning, but there is equally a risk of collapse if we allow it to continue being so unbalanced. Longer-term pruning and focused care will be needed to encourage the other branches to blossom.

7

Masters of the Universe

> I will remember that I didn't make the world, and it doesn't satisfy my equations.
>
> Emanuel Derman and Peter Wilmott, 'The Financial Modelers' Manifesto' (2009)

The use of models in economics, like other modelling endeavours, stems from a wish to control uncertainty about the future. Yet we remain highly uncertain, perhaps more uncertain than ever, about the outcome of modelled economic variables like stock prices or insurance losses over decision-relevant timescales: say, the next ten or twenty years. Events like the 2007–8 global financial crisis have implicated mathematical models as being part of the problem and prompted a rethink of the way that models are used to assess risk by market participants and market regulators.

It's worth recalling that the words 'uncertainty' and 'risk' are used to mean completely different things in different fields. Either way around, they often distinguish between the concepts of *quantifiable probabilities* like the outcome of a coin toss or a repeated experiment and *unquantifiable possibilities* like the chance of a new technology being developed or the political acceptability of a regulatory policy in ten years' time. Many difficulties stem from unquantifiable possibilities being mistaken for quantifiable probabilities. Personally, I prefer to call any unknown outcome 'risk' and to stay in the real world rather than Model Land, usually meaning something that can't be fully quantified. In this, I am broadly in agreement with John Kay and Mervyn King, whose 2020 book, *Radical Uncertainty*, defines risk as the failure of a reference narrative – often negative, sometimes positive and in general not quantifiable. Uncertainty is the lack of certainty about future outcomes that leads to risk.

One of the problems of formal mathematical models is that, in order even to begin writing them down, we have to start by assuming something to be quantifiable. Having done so, we can easily slip further into Model Land, forgetting the difference between model quantities and real ones. Economic and financial models are particularly subject to this form of hubris, perhaps because of the one-dimensional nature of most of the (financial) quantities they aim to describe, or perhaps just due to wishful thinking motivated by what some have called 'physics envy', the desire for economics to be thought of as a quantitative science. So, what is it that models in practice do for the economic and financial worlds? I propose to start with the small scale of financial trading, subject to the decisions of individuals informed by models, and move on to the prospects for regulatory use of models in banking and insurance.

Models making markets

The aim of an individual trader, interacting with the quantified market in financial products, is to use money to make more money by generating a return on financial investment. This is an ideal playground for the mathematician or mathematical modeller, since success is extremely well-defined and the terms of interaction with the market are clearly constrained: if you wish to make a transaction, you must find a counter-party to take the other side of the transaction and thus generate a price for whatever it is you are trading. The existence of a time series of previous data for similar transactions gives you some evidence on which to base decisions. The past data might also give a reason to think that they are created by some underlying generative process which might be modellable and in principle at least some characteristics of its behaviour might be predictable.

Theories of stock behaviour that model prices as a 'random walk' assume that the price movements from day to day, rather than being driven by some predictable process, happen 'at random'. That doesn't mean that they are completely arbitrary, but that they are random draws from some underlying distribution of possible movements. If that's the case, while the individual movements are by definition unpredictable, the overall behaviour can still be expected to follow certain kinds of patterns. Random walks also conveniently happen to be very familiar territory for mathematicians and statistical physicists, since they have applications in a huge range of other fields including statistical mechanics, diffusion in fluids, reaction-diffusion equations of mathematical biology and statistical sampling techniques. It's unsurprising that this simple metaphor travelled from physics into finance. Rather than trying to understand and predict the details, a market participant who does not have a particular interest in any stock or sector can still get involved and still, in many cases, make some money.

On such relatively simple assumptions are built complex mathematical structures of pricing, arcane trading strategies and arbitrage. The random-walk model for prices is extended to the well-known Black–Scholes equation for option pricing: given that a certain stock has in the recent past exhibited movements with certain random properties, we can calculate a 'fair' price to pay for the option to buy that same stock after a period of time has elapsed.

Now, one interesting question is whether the price calculated as being fair is actually the price that is paid for such a contract. Economists Fischer Black and Myron Scholes compared the theoretical predictions of their equation with the actual prices of options traded by a New York broker and found a systematic mispricing in which the sellers of options tended to get a 'better deal' than buyers. A result like this might lead you to think that the model's assumptions are too general and that it simply isn't applicable to the real-world situation. Alternatively, you might think: aha – if these options are mispriced, I can profit from the incorrect assumptions of others in the market.

Arbitrage occurs when prices differ and can be exploited to make a profit. For instance, if the price of a new gadget in a shop at one end of town is £100 but the price in a shop at the other end of town is £150, you could buy ten of the cheaper ones and then stand outside the other shop offering them for £140. In principle, you would make an easy £400, although, to the non-economist, numerous reasons why it might not be such an easy deal will immediately spring to mind. In principle, the possibility of arbitrage ensures that prices do not diverge much within markets, although the same kinds of non-economic considerations apply to markets, allowing for some divergence due to the various frictions of transaction. The efficient market hypothesis formalises the concept that there can be no such free lunch, because if a free lunch were on offer, someone else would have already eaten it. In a sense that's what 'guarantees' that the day-to-day movement cannot be predictable: if it were, someone would predict it and bet in that direction, and the price would move to reduce the predictable signal. Having said that, we will see later that leading proponents of the no-free-lunch idea were quite happy to pick up their own free lunch when they spotted one.

Trekking gadgets from one end of town to the other represents an arbitrage of real world against real world. There is not much to go wrong, except for the possibilities of mugging, arrest, failure to sell your stock or the other shop discounting its own goods to undercut your price. What the implementation of the Black–Scholes equation introduced to mathematical finance was, in a sense, arbitrage between the real world and Model Land. Just as arbitrage in the real world encourages convergence of prices, so the widespread adoption of Black–Scholes option pricing in the 1970s encouraged convergence of the real world with the model. Traders who believed that the Black–Scholes price represented 'the right price' for an option would buy contracts under that price, driving the price up, and then sell contracts over that price, driving it down again. Sociologist Donald MacKenzie famously identified the Black–Scholes model as a performative agent – 'an engine, not a camera'. Rather than being an external representation of market forces, it became in itself a self-fulfilling market force, changing the behaviour of traders and influencing prices. In MacKenzie's language, the traders' use of the model actually 'performed' the model and caused it to be correct. In addition, the 1970s and 1980s were a time when the frictions of transaction were being driven down, directly improving the realism of Black and Scholes's assumptions. Costs and commissions were being removed, communications improved, and electronic and automatic transactions becoming more common. In the decade leading up to the summer of 1987, the theoretical descriptions of the Black–Scholes model were remarkably close to realised market prices for options.

That's not to say there was only one single theoretical prediction. Because the Black–Scholes equation derives option prices as a function of the observed volatility (variability) of previous share prices, we still have the question of how long a period to use to fit the volatility parameter. You could use a long time series from when the stock was originally issued, but this may include data that are completely irrelevant to present performance. Both the value and the volatility change, at least slowly, over time, according to the mood of the market as well as changes in the fortunes of the company. You could use a short time series only for the last few days of data, but this might happen to coincide with a random up- or downtick which is not reflective of the underlying characteristics of the stock, and will fluctuate significantly from day to day. So, in applying the Black–Scholes equation, the trader must also exercise their own judgement about the relevant timescale over which the parameters should be estimated. Going further, the past data are only relevant in so far as they are a guide to the future. What we actually want to plug into the Black–Scholes equation is the *future* volatility: a well-informed trader might have reason to believe that this will differ from the past and incorporate that information into their assessment. In short, like most science, it is an art of applying effective judgement and not solely a mechanical analysis.

The end of a performance

If the traders were 'performing' the Black–Scholes equation into correctness – their assumptions acting as a self-fulfilling prophecy – it was a situation that could not last forever. One of the traps of

Model Land is assuming that the data we have are relevant for the future we expect and that the model can therefore predict. In October 1987, the stock market crashed and market outcomes were not at all the future that was expected, nor were they consistent with any widely used model or prediction based on previously observed data. In particular, it became clear that levels of volatility in stocks are not constant, and they do not always follow convenient log-normally distributed random walks. The events of October 1987 were so far outside the range of plausible outcomes given the models in use at the time that they very effectively falsified the models.

But this crisis of confidence in markets, models and mathematics did not result in the summary ejection of the Black–Scholes equation from the traders' toolbox. Instead, and interestingly, it began to form a framework for discussing model imperfection. Rather than taking the straightforward approach of estimating parameters to determine an outcome (option price), the inverse perspective took the observed option price and back-calculated an 'implied volatility', a quantification of the expectations about the future that were contained in the prices. Then, the implied volatility itself became the unit of comparison and the key information when making decisions, rather than the price itself. As financial economist Riccardo Rebonato nicely put it, the implied volatility became 'the wrong number to plug into the wrong formula to get the right price'.

The Black–Scholes model returns a constant volatility for any strike price of an option (the strike price is the price at which the option can be exercised, i.e., the pre-agreed price at which the buyer has an option to buy or sell). Before 1987, this was enforced (or performed, to use MacKenzie's word) by the traders themselves assuming the correctness of the model. After 1987, rather than a flat line, a 'volatility smile' emerged on plots of implied volatility, tipping upwards for options further away from the central price. That means that the equation is wrong, i.e., that the price no longer supports an assertion that changes will be log-normally distributed in future. Deviations from flatness – a 'smile', 'smirk' or 'skew' – suggest different judgements being made about the way the stock will behave in future.

Having falsified one model, the temptation for any mathematically minded person is usually to construct a new, more complex model that attempts to address the failings of the first; indeed, there are more general forms of the Black–Scholes equation with different random behaviours. Yet, if all models are wrong, there is certainly something to be said for sticking with a simple one that is known to be wrong rather than a complex one that could be wrong in more obscure ways. It is interesting to see how the simpler forms of the Black–Scholes equation have evolved away from a formal predictive framework for pricing and decision-making into something more like an informal, intuitive first guess or benchmark against which deviations are measured and judged.

Trading in Model Land

One of the basic assumptions of trading theory is that some (not all) risks can be hedged away by purchasing different kinds of stocks and options in carefully balanced quantities. For example, the so-called 'delta hedge' is a way to construct a portfolio that is insensitive to small movements in the underlying prices: it doesn't matter whether the market goes up or down, so you can bet on another forecast – such as a convergence in the prices of two stocks – without worrying about what the actual prices are. This insensitivity strictly only applies over infinitesimal time units, so it is a 'dynamic' hedge; in other words, the portfolio must be continually updated as prices change.

Hedge funds exploit these kinds of mathematical tricks to make large bets on specific financial forecasts, arranging the hedging such that some uncertain outcomes are in theory less important to the outcome and reducing sensitivity of the portfolio to these possible variations. In the early 1990s, a fund called Long-Term Capital Management (LTCM) picked up quite a lot of free lunches betting on such things as the likely convergence of price of two different but related holdings, while being delta-hedged against upward or downward movement of the market. Their internal risk management systems were based on assumptions that the observed properties of randomness in the markets over the previous period would more or less continue to hold in future. Under these assumptions, they were genuinely making risk-free returns, subject only to a calculated degree of year-to-year randomness which would average out over the long term. It's clear that the managers of LTCM did fundamentally have a very high degree of confidence in their methods: many invested all of their own personal money into the fund. For several years, exploiting these mathematical principles, they made spectacular returns for investors and huge personal fortunes.

In 1998, LTCM collapsed. Events that it had calculated to be effectively impossible happened, and then happened again and again in quick succession. The individual partners in the fund personally lost a total of $1.9 billion. The fund was bailed out, taken over by a consortium of other banks and gradually wound up over the following few years.

A lot has been written since 1998 about the failure of LTCM, and I won't repeat the usual diatribes against the arrogance and hubris of greedy, overleveraged Nobel-winning economists. Perhaps we can find a more interesting conclusion. What were truly the reasons for the failure of the fund? If the traders had not been so arrogant, there might have been a more effective risk-management programme (as it was, one trader did try to reduce their positions in the summer of 1998). If the fund had not been so heavily leveraged, it would not have been such a catastrophic fall. If it had not made enemies in other institutions, it might have found additional investment earlier and ridden through the crisis to make even more spectacular gains afterwards. If other market players had not copied LTCM's strategies, there might not have been a wave of selling against its position which made everything worse. Many explanations are possible and they all contain some truth. The aspect I want to emphasise here, without downplaying the others, is that the modelled risk drastically underestimated the real-world risk.

Risk management in finance often uses a mathematical construction called Value at Risk (VaR). The VaR can be defined in a few different ways, but in essence it quantifies the largest loss that might occur on a small fraction of occasions, generated from a model based on past data. Then we expect that 99% of the time the loss will not be larger than the VaR. There are a few problems with this simple description. First, the assumption that the past will be a reliable guide to the future is one that has been repeatedly shown to be incorrect. More sophisticated VaR models may price in either expectations about future changes (such as an increase in volatility) or more simply add a conservative margin for error. Second, the VaR has nothing to say about the shape or magnitude of the tail risk itself. The 99th percentile is a useful boundary for what might happen on the worst of the good days, but if a bad day happens, you're on your own. David Einhorn, manager of another hedge fund, described this as 'like an air bag that works all the time, except when you have a car accident'. As such, using the VaR as a risk measure effectively encourages the concentration of risks into the tail above the reported quantile. If you are required to report a 99% VaR, you could, for example, bet on a coin toss *not* coming up heads seven times in a row: you would win 99.2% of the time and the VaR would be zero, no matter how large your bet. And yet if you continue to bet for long enough, the unlikely event will happen, the bet will be lost once and the odds you were offered will suddenly become important again.

Risk managers are supposed to set limits on acceptable trades and volumes, but they face pushback from traders as described by Paul Wilmott, whose caution also heads this chapter: 'If you want to trade but a risk manager says there's too much risk, that you can't – well, there goes your fee.' So the incentives are not clearly aligned in the direction of assessing risk correctly.

The same problem happens in insurance. Let's say Alana and Beth are two insurance brokers offering insurance against a very unlikely event, one that has a 1-in-200-year chance of occurring, but will incur losses of £100,000 per policyholder over a large number of contracts. (Note that I am not talking about house fires here, where any individual house fire might be a 1-in-200-year event but the portfolio consists of many independent and more-or-less uncorrelated risks. This refers to a single large event that will affect all such policyholders.) Alana calculates that the break-even point in terms of expected value would be £500 per year, adds a profit margin and margin for error, and advertises the policy for £650. Meanwhile Beth advertises the same policy for £300. Which policy would you buy? The most likely outcome over the next twenty years is that Beth gets all of the business, the event does not occur, and Alana fails to write any business at all and is sacked, while Beth makes a tidy profit and is promoted. The inevitable outcome in the long run is that Beth's company will eventually be exposed to a disastrous loss that more than negates all of the profits of the preceding years.

Pennies before the bulldozer

This strategy has been termed 'picking up pennies before the bulldozer': there is genuinely free money lying around to be had (at least temporarily) by those who are willing to ignore unpredictable tail risks in the medium or long term.

Who came out better? Alana's career has stalled. Beth has done well, lived the high life and most likely has already retired or even died by the time the tail risk shows up. And if not, of course, Beth still has friends, family, a decent reputation (how could anyone have foreseen such an unlikely event?), and finds a new job very quickly after a suitably contrite period of unemployment.

Which company came off better? Alana's company priced risk 'correctly' and quietly went out of business. Beth's company, underpricing the risk, went from strength to strength. After the catastrophic event occurred, the government deemed that it was unreasonable for policyholders to suffer for the company's failure and bailed out the fund, re-employing Beth (now CFO) and other key employees to wind it down. Beth and partners lost a lot of money on paper, but they promptly set up similar companies to exploit the same strategies and do the same thing again. Those who were less embarrassed by the episode even got after-dinner speaking gigs to discuss their successes and failures. When you are wealthy and privileged, you may lose your wealth but it is much harder to lose your privilege.

This is also, by the way, more or less what happened to the partners at LTCM. I'll talk more about the implications for modellers when their models fail in chapters 9 and 10. For now, the point is that the incentives of risk management are not aligned. There is profit to be had from incorrectly characterising risk, and in particular from underestimating it. Those who correctly estimate significant tail risks may not be recognised or rewarded for doing so. Before the event, tail risks are unknown anyway if they can only be estimated from past data. After the event, there are other things to worry about.

All that is not to say that LTCM were naively taking their models to be reality: in fact, they had a fairly sophisticated risk management system which did not solely rely on a VaR figure calculated assuming that the future would look like the past. And I am sure Alana and Beth could each justify their respective approaches to insurance pricing.

The wider question is whether organisations can be trusted to manage their own risk, given the misalignment of incentives in the long and short term, and the possible misalignment of incentives of businesses versus the people whose money is being managed. If you are only managing your own money, you can take your own risks – all well and good – but in some cases there is also a social interest in the reasonable treatment of financial risk and uncertainty.

If the stake includes, for example, local government pension funds, or affects the compensation of people who have in good faith paid for flood insurance, or becomes large enough to destabilise the whole financial system and require public bailouts, then it is important for the genuine levels of risk to be communicated as well as they feasibly can be.

In doing so, we have to recognise that absolute risk is inherently incalculable. Nassim Taleb gives the delightful example of a casino that can statistically calculate its business risk with high levels of confidence, but this risk is dwarfed by the combination of a tiger mauling a star performer, a contractor attempting to dynamite the casino, an employee failing to file documentation leading to a large fine, and a ransom payment being made for the casino owner's kidnapped daughter. These unmodelled risks, labelled Black Swans by Taleb, are simply not predictable – although we can be pretty sure that *something* could occur.

Regulation using models

The true incalculability of risks poses a problem, therefore, for the regulator, who would like to be able to set a defined limit on the amount of risk shouldered *by society* when private enterprises incur losses that are too great for them to bear personally. How do you stop people piling in front of the bulldozer?

Risk management in banking has been led by the Basel Committee on Banking Supervision, which issued its first set of guidelines (30 pages) in 1988, a second set (347 pages) in 2004 and a third set (616 pages) in 2010. The expansion in complexity is concerning in itself, moving from a handful of asset classes to disaggregation into many different kinds of exposures, all with individual risk weightings. In addition, the banks have been allowed to use their own internal risk models rather than a mandated calculation. Can a regulator – or for that matter an investor – understand the level of risk that a bank is taking on when it provides only a brief description of its modelling methodology and a final answer? Andy Haldane, then a director of financial stability at the Bank of England (the UK's central bank), set out in a 2012 speech his position that a reduction in the complexity of banking regulation would be desirable. Many other people seem to agree, but concrete proposals for actually achieving that are less easy to find.

And they are swimming against a tide. Given the ever-increasing complexity of financial instruments and trading algorithms, the complexity of internal risk models must be increasing too. I expect Basel IV will have to account for internal risk models estimating parameters using an unexplainable AI-based method. Robert Merton, one of the directors of LTCM, was quoted by the *New York*

Times in 1999 as saying, 'The solution is not to go back to the old, simple methods. That never works. You can't go back. The world has changed. And the solution is greater complexity.'

It is also possible that the risk models contribute to risk itself. Setting quantified limits on risk exposure encourages institutions to max out their risk-seeking within that limit, even when it might not be a good idea to do so. Almost by definition, in a financial crisis the models fail, because the character of market movements is quite different from the previous variability on which the models are calibrated. And the models are active players in the market, not passive observers. If the internal models used by a big bank were to predict a financial crisis and take rapid risk-limiting action, it would almost certainly precipitate the crisis rather than avoiding it.

Insurance is a slightly simpler case, at least where the insured risks are physical. In principle, you can diversify exposure to life insurance risk in Europe by taking on other risks such as life insurance in another region or weather hazards, which you expect to be only moderately correlated with the first risk. If you can find a set of completely uncorrelated risks, you can reduce the expected overall volatility of your portfolio by combining a little bit of everything. This is the principle of any insurance: by pooling risks, the shared pot remains reasonably stable while paying out to mitigate what would be catastrophic losses for individuals.

The difficulty is assessing exactly how independent different risks actually are. Events that are generally uncorrelated most of the time may become correlated in the face of large events. Before 11 September 2001, commercial property in New York would not have seemed much correlated with airline business continuity. Before the Covid-19 pandemic, life insurance in the US would not have seemed much correlated with holiday insurance in Europe. Most of the time, it is reasonable and justifiable to model these kinds of events as being uncorrelated, but just occasionally something happens that ties them together, and this kind of large event is where insurers are vulnerable. The EU Directive called Solvency II is somewhat like the Basel banking requirements in that it specifies how insurers should calculate levels of capital to hold in preparation for possible large events.

It uses something a bit like the VaR approach, requiring insurers to calculate a 1-in-200-year event (99.5% VaR). Doing that requires a model that can be calibrated on past data but will again have to make assumptions about how closely the future can be expected to resemble the past. A 1-in-200-year flood loss event, for example, is difficult to calculate for various reasons. First, we may not have anywhere near 200 years' worth of past flood data, so the 1-in-200 level is found by extrapolation. Second, even if we had 200+ years of past data, we are unsure whether the conditions that generate flood losses have remained the same: perhaps flood barriers have been erected; perhaps a new development has been built on the flood plain; perhaps agricultural practices upstream have changed; perhaps extreme rainfall events have become more common. Our simple statistical model of an extreme flood will have to change to take all this into account. Given all of those factors, our calculation of a 1-in-200-year flood event will probably come with a considerable level of uncertainty, and that's before we start worrying about correcting for returns on investment, inflation or other changes in valuation.

Model risk

So, for both banks and insurers, on top of the VaR assessed by the models, there is a 'model risk' – the possibility of a discrepancy between what the model forecasts and what you actually observe. Model risk can rear up very quickly when extreme events happen. The so-called twenty-five standard deviation moves noted by Goldman Sachs during the financial crisis represent an enormous model risk (or model failure), even though the same model would have presumably been reliable and profitable for a long time before that. Emergent correlations in insurance are another kind of model risk. These are the kinds of events that bankrupt large companies and destabilise international institutions.

How can a regulatory body take a coherent position on model risk, when by definition it is the contribution to risk that is unmodellable? There isn't going to be a way to forecast exactly what form the next huge loss event might take, or when it will happen, but we can't even estimate meaningfully the probability of it occurring in the next five years. Nor is it reasonable without dramatic cultural change to expect insurers and bankers to hold capital to cover all possible eventualities, because that would put them out of business in the very short term.

One partial answer is stress tests. The most recent insurance stress tests include running scenarios of concurrent windstorm and flooding, or a large cyberattack (but not a cyberattack officially sanctioned by another nation, because that would count as an act of war and therefore be excluded from policies). You can also imagine that the California 'Big One' earthquake has been well modelled, as well as major hurricanes hitting large population centres. Those who ran simulations of a viral pandemic in the last few years have had an interesting chance to check up on their assumptions since 2020. Pandemic appears not to have been officially used as a stress-test scenario by regulatory bodies, but health insurers at least have certainly had pandemic influenza as a known major possibility for years. Even where scenarios exist, the many and varied downstream impacts of the pandemic that are emerging, from supply-chain disruption and Long Covid to accelerated digitisation and the 'Great Resignation', are potentially major economic changes that will not have been factored in.

Stress tests in banking are less colourful but include scenarios for economic downturn, changes in interest rates or the failure of another major bank. After the Covid-19 pandemic began, the Bank of England conducted stress tests for the projected downturn and then a set of 'reverse stress tests' that assessed how bad things would have to get to jeopardise UK banks' ability to function. The reported result of these reverse stress tests in August 2020 was that things would have to be about twice as bad as their central projection to start to cause structural failures. The 2021 stress test looked at a 'double-dip recession' including ongoing severe economic challenges such as UK house prices falling by a third and UK and global GDP each falling by about a tenth. The use of these stress tests, in theory, allows banks and the regulator to identify areas for concern and to strengthen their positions in order to be able to continue to provide essential services.

The dog and the frisbee

In the speech I mentioned earlier, central bank economist Andy Haldane compared bank regulation to a dog catching a frisbee, saying that complex systems are best governed by simple heuristics rather than by trying to optimise a complex response strategy.

> Modern finance is complex, perhaps too complex. Regulation of modern finance is complex, almost certainly too complex. That configuration spells trouble. As you do not fight fire with fire, you do not fight complexity with complexity. Because complexity generates uncertainty, not risk, it requires a regulatory response grounded in simplicity, not complexity.

Again, he is drawing the distinction between quantifiable possible futures (risk) and unquantifiable ones (uncertainty).

In that environment, as a dog trying to catch a frisbee, what is it that more models can do for us? First, we should remember that the dog has to learn how to catch a frisbee and will get better at it with practice. Maybe you remember playing 'Catch' as a child and someone shouting at you to 'keep your eyes on the ball!' It's a very good start to have a learned model that says that you can do better by continually observing the position of the object than by taking a glance and then running to where you think it will land.

But there is a problem of misaligned incentives. The financial regulation dog is not just trying to catch a frisbee, it is trying to catch multiple moving drones that are being deliberately piloted to avoid capture. The aim of the regulator is to maintain the stability of the overall system. The aim of the regulated participants in the market is to make money for themselves as individuals and their companies as a whole. The kinds of limits imposed by the regulator in the short term work against opportunities for profit. Alan Greenspan, ex-chair of the US Federal Reserve, said after the financial crisis, 'Those of us who have looked to the self-interest of lending institutions to protect shareholders' equity, myself included, are in a state of shocked disbelief.' It would be convenient to think that a Darwinian 'survival of the fittest' approach to markets would enforce this kind of self-interest, but it fails due to the immediate feedback of competitive pricing outweighing the long-term feedback of system stability. Self-interest in principle should include longer-term sustainability of the system, but in practice those market participants who do not price in longer-term sustainability can be more competitive and put the rest out of business. The trouble is that there are a very large number of pennies in front of the bulldozer and they can make a lot of people very rich very quickly.

Economist Joseph Stiglitz, writing in 2018 about the dynamic-stochastic general equilibrium (DSGE) models used as tools for macroeconomic analysis and forecasting, identifies the same trend towards complexity:

> As many users of DSGE models have become aware of one or more of the weaknesses of these models, they have 'broadened' the model, typically in an ad hoc manner. There has ensued a Ptolemaic attempt to incorporate some feature or another that seems important that had previously been left out of the model. The result is that the models lose whatever elegance they might have had and claims that they are based on solid microfoundations are weakened, as is confidence in the analyses of policies relying on them. The resulting complexity often makes it even more difficult to interpret what is really going on.

It is sometimes said that 'it takes a model to beat a model', with the implied meaning that one should not criticise someone else's best attempt unless you can make a better job of it yourself. I suppose we cannot yet tell whether there is some 'heliocentric' model that will explain and predict the global economy much better than any previous Ptolemaic attempt, although it seems unlikely. But the rationale of needing a model to beat a model is clearly flawed, if our view is one of adequacy-for-purpose rather than theoretical elegance. Central banker Mervyn King and economic writer John Kay have argued as such, saying, 'If a model manifestly fails to answer the problem to which it is addressed, it should be put back in the toolbox . . . It is not necessary to have an alternative tool available to know that the plumber who arrives armed only with a screwdriver is not the tradesman we need.' If the frisbee is thrown over a brick wall, will the dog crash into the wall in pursuit, or will it instead stop, rethink and go around through the gate?

Carolina Alves and Ingrid Harvold Kvangraven have examined the turn of economics towards modelling. They say that

> an increasing number of economists think of themselves as modellers, 'simplifying' reality through models and invoking the necessary assumptions regarding equilibrium, representative agents, and optimisation . . . A consequence of this turn is that economists find it increasingly difficult to imagine different ways to do economics.

The attachment to a certain way of working means that failures, instead of prompting a rethink of the model, result in a move towards more complexity and elaboration. The DSGE models that failed to predict the financial crisis were not adequate for purpose: as Stiglitz also points out, given their working assumptions, 'a crisis of that form and magnitude simply couldn't occur'. At what point do you stop adding epicycles and move instead to a new model?

Learning from samples of one or fewer

Instead of a physical time series, adequately represented by some mathematical equation in which the value at each time point is some fiendishly complex function of the previous ones and other information, Elie Ayache has posited a view of the markets that is much less amenable to quantitative analysis: 'all that we have is a succession of prices, which are in fact unique outcomes of contingent claims between traders, and whose quantitative form does not imply that they are parts of any coherent numerical series.'

Effectively, he says, we are learning from a past sample whose relevance to the future we cannot clearly define. The past is ungraspable just as the future is unknowable, so that in order to experience either past or future as anything other than a succession of measurements, we have to impose some structure by means of a simplifying model. But when we do so, we create a shape for our own thinking that may be useful and may equally be misleading. Where the modeller endows their model with their own values, priorities and blind spots, the model then reflects those values, priorities and blind spots back, both in terms of how the system works and in terms of the kinds of interventions one could make. Stiglitz noted exactly this, that 'our models do affect how we think'.

Paul Wilmott and Emanuel Derman included in their Modeller's Hippocratic Oath the following: 'I will remember that I didn't make the world, and it doesn't satisfy my equations.' While taking into account known uncertainties (via models), we also need to structure our thinking so that it is not completely blindsided by unknown or unknowable uncertainties. A classic work in management theory, James March, Lee Sproull and Michal Tamuz's 'Learning from Samples of One or Fewer' describes some of the more complex relations between quantitative inference, qualitative thinking and strategies for making decisions outside Model Land. It emphasises organisational accounts of dealing with uncertainty in situations like avoiding nuclear accidents, where one hopes to have no data at all to go on, demonstrating the necessity of imaginative construction of counterfactuals and hypotheticals for these scenarios, supported but not limited by quantitative inference from available data.

Becoming too attached to a single kind of model, in a single region of Model Land, leaves you vulnerable to excluding the possibilities that might be generated by this kind of imaginative exploration. The article describes how the expert judgement embedded in imaginative counterfactuals can be a first-order contributor to risk assessment. Before the space shuttle Challenger exploded soon after launch in January 1986,

> the spacecraft flew a series of successful missions despite its faulty O-rings. Some engineers interpreted the indications of O-ring problems during these early flights as symptoms that past successes had been relatively lucky draws from a distribution in which the probability of disaster was relatively high. Others, including some key personnel in NASA, considered these estimates of danger as exaggerated because, in the realized history, the system had been robust enough to tolerate such problems.

On the face of it, and with reference to the data, either scenario is feasible. Only the application of expert judgement gets us out of Model Land and back into the real world.

We should be worried

The use of models in economic and financial situations deserves a longer exposition than I have space for here. But what I hope I have shown in these few brief examples is that, in this inherently social sphere, mathematical modelling cannot be expected to work in the way that it might naively be hoped to. The gap between Model Land and real world is not necessarily always large but it varies over time and is unpredictable by analysis of previous data. Despite the allure of quantification, unquantified and unquantifiable uncertainties abound, and the objects of study themselves are directly influenced by the models we put to work on them.

In this fast-moving field, models are useful for a time – sometimes extremely useful – and then fail dramatically, rather than gradually departing from a resemblance to reality. The interconnectedness of models and markets poses a danger: we cannot know when the next financial crisis will come, but it is sure to come, and the trend is towards financial crises with larger and larger systemic effects.

Regulators seek to reduce the potential risk to society, but are reliant on the same models to do so and are chasing complexity with complexity, a top-heavy burden that itself may have the unintended consequence of increasing rather than reducing real risks.

Those who make a career facing day-to-day decisions up close to this uncertainty have generally learned strategies for coping with it. Some of that involves humility, remaining sceptical of any optimisation and adding in margins for error. And some of it quite the opposite: it may involve strategically mispricing risk, safe in the tacit knowledge that this is often a good gamble and that a zero in a bank account doesn't need to mean a permanent exit from the game. It might include using super-fast models to fleece less-informed players with smaller computers rather than making judgements about the fundamentals at all. It might include betting against those quantitatively seduced players who haven't yet learned not to take their models literally. These strategies are not conducive to stability. The continuing increase in computing power and the entry of machine-learning methods into this environment further tip the balance towards variability and unpredictability. We should be worried.

Are there solutions? None that everybody agrees on, yet. Regulators tend not to intervene actively in price setting, fearing that 'market distortions' (as if there were such a thing as an undistorted market) would cause further negative effects. Instead, post-2008 reforms include commitment to the Basel capital requirements, determined by risk models for different assets, and the 'countercyclical buffer' which aims to build up capital in calm periods which can be drawn down when the storm comes. Roman Frydman and Michael Goldberg propose a set of 'guidance ranges' and other dampening mechanisms to reduce the rapidity of asset-price movements beyond reasonable excursions. However, as Ben Bernanke put it in 2002, before he became chair of the Federal Reserve, one cannot assume that the regulator will necessarily be better able to judge the reasonableness of a price than the combined judgement of lots of financial professionals. Nassim Taleb's strategy of 'antifragility' – creating systems that are stronger after shocks – suggests allowing banks to fail so that the banking ecosystem can be more resilient, which perhaps implies a more Icelandic strategy of bailing out citizens rather than institutions. Iceland's post-crisis experience, however, could hardly be described as attractive. A more pedestrian but effective way of dealing with risk is by engineering the system for redundancy and reducing connectedness so that shocks, when they occur, cannot propagate through the whole system and bring all of it down. This, however, is very much counter to the actual direction of travel of the financial system in recent decades, and it introduces 'inefficiencies' that are anathema to the principles of modern finance. There has been some talk about splitting up banks so that none is 'too big to fail', but the list of Systemically Important Banks stands at thirty institutions in 2021, up from twenty-nine when the list was first published in 2011. We are still living in Model Land and trying to optimise. As such, we are systemically unable to anticipate shocks, because the act of truly contemplating one might be enough to precipitate it. The only forecasters who could offer a guarantee of doing well are bad actors who might manufacture a crisis deliberately. Ensuring that single agents (individuals, algorithms or institutions) do not have the power to do this, then, is certainly a priority.

In the present situation, unknowability and the consequent inability to optimise mean that there is no silver bullet that can solve either the problems of individual long-term market participants or those of the regulator. So I agree with Kay and King's conclusion that we must embrace uncertainty. Where I depart from their analysis is in the question of what to do next. If we are to make sufficient allowance for the unexpected, we can only do so outside Model Land. There has been a trend away from charismatic individuals making big decisions based on personal instinct, and towards emotionless computers making big decisions based on data and models. Having read the foregoing chapters, you will not be surprised that I think a strategy for appropriate long-term management of the markets is not to be found in either extreme, but in a considered synthesis of human judgement with mathematical support. The same problems of accountability arise, as we saw in the last chapter. Who are to be the humans in the loop? What are they acting to achieve? How can the rest of society have confidence that they are acting in our best interests as well as their own? Again, these are questions that extend beyond Model Land and into the real world. As far as I'm concerned, though, these are critical questions for finance and economics in the twenty-first century.

8

The Atmosphere is Complicated

> Perhaps some day in the dim future it will be possible to advance the computations faster than the weather advances and at a cost less than the saving to mankind due to the information gained. But that is a dream.
>
> Lewis Fry Richardson, *Weather Prediction by Numerical Process* (1922)

Models of the physical world have some advantages over models of social phenomena. The physicist generally starts with an expectation that there is something real to be modelled and that the model can be more of a camera than an engine. In many cases, that makes a very successful framework for modelling and prediction, and examples of this success fill high-school physics textbooks. But complex systems like the weather have mathematical properties which mean that we have to take a more careful approach to prediction and the evaluation of model performance. And when models interface with social and political decisions, as in the case of climate modelling, then we also need to start thinking about the social and political dimensions of the models themselves.

Richardson's dream

The first numerical weather forecast was both a stunning success and an abject failure. Lewis Fry Richardson, a British Quaker and pacifist, resigned his post at the UK Met Office to spend the First World War as an ambulance driver in France. A mathematician and physicist, he had a career astonishing even by the standards of late-Victorian scientists. In particular, Richardson developed the basic idea of modern weather forecasting: to divide up the landscape into a chequerboard of grid cells, writing a set of nonlinear partial differential equations for each point that expressed the balance of atmospheric variables, then performing a series of tedious calculations that advanced one step at a time, with the result for each cell remaining consistent with its neighbours and its previous state.

To demonstrate and test the method, Richardson laid out a series of forms with quantities and ordered calculations, essentially prefiguring the spreadsheet and the algorithm. He took two years to perform one step of his forecast by hand with a 25cm slide rule, laboriously tabulating each component of each term of each equation to three significant figures and generating a forecast for six hours ahead for a town in Bavaria. He wrote that his calculations

> were worked out in France in the intervals of transporting wounded in 1916–1918. During the Battle of Champagne in April 1917 the working copy was sent to the rear, where it became lost, to be re-discovered some months later under a heap of coal.

After such a heroic effort, carried out between other heroic efforts and in the challenging surroundings of 'a heap of hay in a cold rest billet', the results were terrible. There were unreasonably steep pressure gradients and the forecast changes in temperature over the six-hour period were too great to be believable. This was discouraging both for Richardson and, at least for a time, for the wider weather-forecasting community. Returning from the war, Richardson published his method in *Weather Prediction by Numerical Process* in 1922, but the failure of the demonstration and the huge human resource that would be required to do it at scale initially worked against further exploration of the concept. In the book, he also suggests a way of ordering the computers (which were of course humans with slide rules, not computers as we know them today) by parallelising their calculations in this extraordinary vision:

> Imagine a large hall like a theatre, except that the circles and galleries go right round through the space usually occupied by the stage. The walls of this chamber are painted to form a map of the globe. The ceiling represents the north polar regions, England is in the gallery, the tropics in the upper circle, Australia in the dress circle and the Antarctic in the pit. A myriad computers are at work upon the weather of the part of the map where each sits, but each computer attends only to one equation or part of an equation.

Richardson was well ahead of his time: conceptually, this is very much how numerical weather prediction was later implemented, but using punch-card machines rather than a room full of people. He retired in 1943 to pursue his studies in peace and security, and in March 1950 the first computational weather forecast was performed, using a different numerical method but clearly inspired by Richardson's dream. Sadly, he died in 1953 before the age of computing really took off, but from these first steps we now have access to weather forecasts at our fingertips, calculated on computing resources beyond Richardson's wildest imaginings.

What went wrong with his original forecast? Recent work has re-examined his calculations and found there to be only trivial errors in the integration he spent so long calculating by hand. The error, essentially, was in the set-up of the problem: the initial conditions were out of balance, resulting in immediate compensatory swings which dwarfed the predictive signal. Peter Lynch has recomputed Richardson's original forecast with smoothed initial data, repeating and extending the algorithm to show that the method itself would have provided a genuinely useful forecast – an incredible vindication and a fitting conclusion to Richardson's extraordinary story.

The Butterfly Effect

Short-term numerical weather forecasts have improved dramatically since their arrival in the 1950s and, with the continued relentless increase in computing power and assimilation of new observations, performance is still improving today. The UK Met Office boasts on its website that a four-day-ahead weather forecast today is as good as a one-day-ahead weather forecast was thirty years ago. Similar improvements can be seen across the board for different climate variables and different modelling centres, although the absolute level of skill varies a lot.

Yet despite all this progress, we are still limited. The forecast from your favourite weather provider probably only goes ten days into the future, even though it's based on information from some of the biggest supercomputers available, and that's because the information beyond that point is increasingly uninformative. In most areas, if a weather forecast provides daily detail further than about two weeks ahead, you can safely scroll past without losing any information. It's filler for advertising space, not a useful forecast.

The atmosphere is difficult to predict because it is both complicated and complex. It is complicated in that it encompasses many interconnected physical aspects such as air movement, pressure, humidity, gravitational forces, rotation, evaporation and condensation, radiation, diffusion, convection and friction. And it is also a complex system, one in which larger-scale behaviours and properties like hurricanes, the jet stream and El Niño emerge from the combination of these local physical processes acting together. On the smaller scale, thermodynamics dominates and we can write and solve equations balancing pressure and temperature gradients. On the larger scale, dynamics or large-scale movements of air masses and their interactions with land and water are more important. This is what makes UK winters, for example, much warmer and wetter than those of central Canada at the same latitude, despite receiving the same solar radiation.

As a complex dynamical system, the weather is unpredictable in interesting ways. One source of unpredictability is the Butterfly Effect:

the way that very small changes to the initial conditions can result in completely different outcomes after a relatively short period of time. Edward Lorenz described in his 1993 book, *The Essence of Chaos*, the sudden realisation he came to while working on a long calculation:

> At one point I decided to repeat some of the computations in order to examine what was happening in greater detail. I stopped the computer, typed in a line of numbers that it had printed out a while earlier, and set it running again. I went down the hall for a cup of coffee and returned after about an hour, during which time the computer had simulated about two months of weather. The numbers being printed were nothing like the old ones.

What Lorenz had done when he restarted the calculation was to replace the original numbers with very slightly different rounded-off ones. This tiny error increased a little, then a little more, and 'more or less steadily doubled in size every four days or so, until all resemblance with the original output disappeared somewhere in the second month'.

Although it's now a well-known concept, it isn't at all intuitive: one might think that a small error in the initial condition would result in similarly small errors later, perhaps with some proportional increase over time. Indeed, this is the case in many systems with which we are familiar. If you throw a ball at a hoop and your aim is just slightly wrong, you miss the hoop only by a small distance. Yet what Lorenz had found was that, for complex systems, small errors in initial conditions could increase proportionately, growing out of control until all useful information was lost. This is partly why the exponential increase in computing power has resulted in only linear improvements to the lead time of useful weather forecasts.

The problems caused by the Butterfly Effect have an obvious solution. Just measure the initial conditions more accurately and you'll get a more accurate forecast. As the importance of initialisation has become more and more clear, weather-observing systems have been massively improved over the last century, now assimilating near-real-time data from satellites, aeroplanes, ships, weather stations, radiosonde balloons, radar systems and more.

There is also a slightly less obvious solution to the Butterfly Effect, one that has prompted huge changes in the way that weather forecasts are made and communicated. If we can't pin down the initial condition perfectly, instead of taking a single guess, we could take several guesses and use the resulting scatter or 'ensemble' of predictions to give an idea of the range of possible outcomes. In other fields such as finance this kind of approach is sometimes called a 'Monte Carlo simulation'. Major weather-forecasting centres now use an 'ensemble prediction system' which does exactly this, presenting a set of different forecasts that arise from slightly different initial conditions which are all consistent with the observations. Public forecasts do not often show all the options, but they might acknowledge the uncertainty by presenting weather probabilities rather than an absolute prediction. The most widely recognisable examples of ensemble weather prediction are probably the 'spaghetti plots' showing possible hurricane tracks, which are sometimes merged into a single 'cone of uncertainty' showing the area thought to be at risk.

Now, if we had perfect models (despite imperfect measurement), this would be a solution to the problem posed by the Butterfly Effect. We can even make it probabilistic. If the initial distribution of ensemble members is reflective of our confidence in the initial conditions, then the final distribution of ensemble members would be a reliable probability distribution about the outcome (top-right circle in Figure 3). We could form a 90% confidence interval, for instance, and expect that the true outcome would fall within that interval 90% of the time. Randomness is a solution to chaos. The accuracy of measurement still limits the timescale over which you can usefully predict (days to weeks, in the case of the weather forecast, depending on your use case), but if we had perfect models and big computers, we could almost always predict the range in which the future outcome will fall, though not its exact value.

The Hawkmoth Effect

Unfortunately for us, and for the weather forecast, that is a very big 'if', because we don't just have errors in our measurements of today's weather, we also have errors in our representations of the physics of weather: that is to say, errors in the structure of our models. That might be due to neglected factors, or 'unknown unknowns', or more prosaically just due to numerical integration of continuous processes. For some simpler problems, again, we can be in the more intuitive situation that a small error in the model representation would result in similarly small errors later, again perhaps with some proportional increase over time. If I write down a slightly wrong equation for the trajectory of a basketball – say I forget to account for air resistance – then I expect that my prediction will be only slightly wrong. But for complex and nonlinear systems like the weather, small model errors can instead result in large prediction errors even over short timescales. This is the Hawkmoth Effect.

When the Butterfly Effect strikes, our forecasts start off accurate and then become imprecise, but they do not become misleading. When the Hawkmoth Effect strikes, our forecasts start off accurate, then they can become misleading (see Figure 3). Imprecision is not a sin, and it need not cause bad decisions as long as we know to expect it. When a forecast is misleading, however, it can cause us to do the wrong thing in mistaken expectation that we know what will happen.

The Hawkmoth Effect is analogous to the Butterfly Effect, but rather than sensitivity to the initial condition, it describes sensitivity to the *model structure*. You might think we could solve this problem with a slightly different Monte Carlo approach: taking a lot of alternative models and using the output of this multi-model ensemble to generate a probability distribution of potential outcomes in the real world. Great idea – but wait! The initial conditions were real-world measurements and there were only a finite number of dimensions in which they might be wrong: the position of a basketball can only be incorrect by being too far up, down, left, right, forward and/or back. We can systematically try a range of possibilities that encompass the actual position of the ball. That's why the Butterfly Effect is solvable: we can know that the true outcome is somewhere within the range of predicted outcomes.

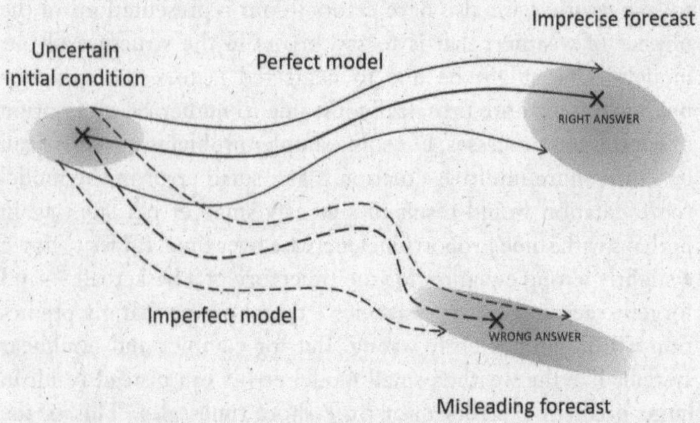

Figure 3: The solution to the Butterfly Effect is to run a perfect model with lots of different initial conditions (solid lines). The fuzzy circle on the left shows the range of measurement uncertainty around the correct initial condition, and the fuzzy circle at top right shows the resulting imprecise forecast. Doing this with a perfect model (solid lines) will produce a range that contains the correct answer. But when models are imperfect (dashed lines), the range of uncertainty generated by this kind of procedure may well not contain the correct answer. The Hawkmoth Effect says that even when the models are in some sense 'only slightly wrong', the divergence of dashed lines from solid ones can still be significant.

But alternative model *structures* exist in Model Land, not in the real world. There is (hopefully) only one actual effect of gravity, but there are infinitely many possible ways it might be slightly different and we can only write down the ones that we happen to be able to think of. We cannot know whether any of them are correct, and because there are infinitely many dimensions, we cannot be confident that the truth is encompassed by the range of those we have specified.

Then we have to ask ourselves what it is about alternative possible models that tells us anything at all. Why should we expect those different Model Lands to contain any information about reality? If one hypothetical simplified universe wasn't good enough, why should a mash-up of other hypothetical simplified parallel universes be any better? A structural sensitivity analysis of a model tells us about the properties of the model, not about the real world.

Eric Winsberg, a philosopher of science, has written extensively about the Hawkmoth Effect, claiming that it is unnecessary to consider it in the context of climate models. In his 2018 book on the philosophy of climate science, he asks:

> What reason is there to think that climate scientists make mistakes about the order of the polynomials their models should have? Or that they sometimes write down exponential functions when they should have written down sinusoidal ones? Corollarily, what reason is there to think that small model errors . . . will . . . produce deviations on such short timescales?

These are common concerns. Winsberg answers his own rhetorical questions, however, in the final paragraphs of the book itself, where he notes the possibility of abrupt and disruptive changes that are not modelled adequately and that could happen over a very short period of time. In failing to model 'known unknowns' like methane release or ecosystem collapse, climate modellers are indeed writing a zero-order polynomial when we know for sure that the truth is something more complex.

We know what the timescale of applicability for weather models is: a few days to weeks, determined by the timescale on which we find major divergence in the range of forecasts. This much is obvious because we can see the deterioration of usefulness for any particular application by direct observation, partly due to the initial condition measurements and partly due to imperfection in weather models. But for climate models we have no such intuitive understanding of the timescale upon which they might become uninformative or even misinformative. How do we know what the informative timescale is for a climate model? In practice, climate models are typically run up to 2100, but this is an arbitrary cut-off at a round number, not one based on an assessment of predictability constraints. It would be useful to know if the effects of non-modelled processes might become important on timescales of decades, or if they can be safely neglected for a couple of hundred years.

Such an analysis would give us greater confidence in the probabilistic predictions of future climate. By the way, the existence of the Hawkmoth Effect is not a reason to doubt the basic physics of climate change, and it is not a reason to doubt that the characteristics of projections made by climate models are real possibilities. It is only a reason to be cautious in interpreting the detailed projections of climate models, and especially cautious in assigning probabilities to outcomes. Climate change is happening, it will have serious consequences and we must decide what to do about it in the absence of perfect information. If the presence of the Hawkmoth Effect serves to widen somewhat the ranges of plausible outcomes, it also supports a stance of humility about our ability to model, predict and adapt to changes, and therefore speaks in favour of greater efforts in climate change mitigation by reduction of emissions.

Not noise, but music

Fundamentally, the problem is one of complexity. We are interested in learning about a complex, many-dimensional dynamical system (the climate) and we create another complex, many-dimensional dynamical system (a climate model) as a proxy to learn about it. But when you subtract one complex, many-dimensional dynamical system from another, you do not get simple random noise. If you did, you could model that noise statistically and gain an understanding of the average error in different circumstances. Instead, what you get when you subtract climate from climate-model is a third complex, many-dimensional dynamical system, which has its own interesting structures, correlations, patterns and behaviours. As put by eminent meteorologist and climatologist Brian Hoskins, it is not noise, but music.

This distinction between random statistical noise and non-random dynamical 'music' is of interest because of the different methodological approaches these two perspectives suggest. If you think your errors are random and therefore in principle not understandable, you will incline towards methods that seek to reduce uncertainty by averaging out the randomness, waiting for a 'signal to emerge from the noise' and so on. The ensemble methods described above take this statistical approach and, as we have seen, they work very well for genuinely random noise (measurement error of initial conditions) but poorly for dynamical music (errors in the specification of model structure). If you think your errors are dynamical in nature, you might seek instead to improve the quality of the underlying model, using the structure of the error as an indication of what might be going wrong and perhaps how to fix it.

Disappointingly, the nuances of complexity also make this strategy somewhat difficult. With a simple model and a small number of parameters, the structure of error is relatively easy to read. Say you model a very simple Earth system and you find that the distribution of residual errors is a nice smooth sine wave with a period of one year, then you can easily realise: Aha! I should include a seasonal cycle. But as a model increases in complexity, so too the structure of error becomes more complex. Subtracting one complex, many-dimensional dynamical system from another yields residuals that are harder to interpret because they do not themselves correspond neatly to physical processes, but to the difference between one and another.

We can see this in relation to the development of weather models, for instance. The European Centre for Medium-Range Weather Forecasting (ECMWF) is one of the world's leading weather-forecasting centres and has a constant programme of model development, with major upgrades introduced regularly. When it introduces an upgrade, it typically improves performance in some aspects and makes others worse. The effects of the 47r3 update made in October 2021, for example, were illustrated by a colourful scorecard showing the level of change for each variable in different geographical regions. The update improved the representation of water-cycle processes like clouds, convection and rainfall, and as such it made a significant improvement to the score for total precipitation, but with small deteriorations in other variables. As the ECMWF said in its press release, 'With such a major physics change, it is inevitable that there are some degradations'.

This is a common problem when models reach a high level of complexity. After deciding on the structure and representations to be used, there is a process of tuning or calibration, during which you have to choose the best set of parameter values. One group of researchers described 'the art and science of climate model tuning', noting that all tuning methods involve subjective choices about which metrics are the most important ones. Rather than achieving universal improvements, tuning a model to get the best performance in one aspect tends to degrade performance in other respects. And there is disagreement about what constitutes a legitimate target of tuning. We hope, for example, that climate models will make good predictions of the global average temperature. In a 2014 survey, about a third of modellers said that this meant we should calibrate models to reproduce observed twentieth-century warming well and thereby improve the likelihood that the processes underlying the observed warming have been represented correctly. Another third said that we should not calibrate directly to twentieth-century warming, that to do so is an unjustifiable short cut, and the aim should instead be to calibrate models based on more fundamental physics and then see what level of warming comes out as an emergent property. Both of these are reasonable positions: calibrating a complex model is certainly an art as well as a science.

From weather to climate

Climate projections are based on weather predictions, although they do not work in exactly the same way. Where weather models are used to predict the detail of atmospheric patterns over time and their predictive skill decays on the order of weeks, climate models are used to predict larger-scale changes in statistical patterns rather than the exact evolution of the system. Robert Heinlein's adage that 'climate is what you expect and weather is what you get' expresses this well: even if we cannot predict exactly what we will get from day to day in the 2080s, we hope to be able to predict the way in which weather patterns will change. This follows from the original definitions of 'climate' which contrasted the typical weather patterns found in different geographical regions by building up a statistical picture over many years of observation.

Of course, as 'sceptics' are wont to point out and scientists to agree, climate has always been changing. With or without humans, this complex dynamical system has internal processes that change on all timescales, from gusts of wind to thunderstorms, cold spells, the seasonal cycle, El Niño, the Ice Ages and continental drift. So it is a challenge to decide exactly where the cut-off should be placed. If we define climate on too short a timescale – say, three years – we might just so happen to include three years that are rather like each other and not so much like the next five years. If we define climate on too long a timescale – say, 300 years – we would be including data (which we may not even have available) from a quite different climatic regime. And if we expect climate to change rapidly in the future, a definition over too many years simply won't be able to keep up. In practice, climate scientists have often used a thirty-year sample size, which represents a convenient but arbitrary trade-off between these two ends of a spectrum.

Thirty years is also a length of time over which human memory can perceive regularity and change. If humans had a lifespan of only one year, or of 10,000 years, our idea of climate might be quite different. The idea of what the weather 'should' be like in a certain place at a certain time of year is anchored to personal experience, whether that is your own experience of sowing seeds at a time of rain or a storybook idea that it often snows at Christmas. As geographer Mike Hulme has described, 'climate' is really a constructed idea that helps to stabilise cultural experiences of changeable weather. As such, it also takes on cultural forms that are appropriate to the cultural context. The scientific idea of meteorological climate corresponds to a modern Western cultural separation of human from nature, but other conceptualisations of the human relationship with weather can be more integrated. Marshall Islanders of the Western Pacific, for instance, translate 'climate' as *mejatoto*, but this term encompasses both physical and social surroundings including social relationships. Historical accounts of European witchcraft speak of the powers of witches to induce or prevent rainfall or extreme weather, and weather events have often been taken as supernatural omens. Even in the most 'scientific' and 'modern' cultures, we can still find views like those of David Silvester, a local councillor from Oxfordshire in the UK, who stated in a letter to a local newspaper in 2014 that 'Since the passage of the Marriage (Same Sex Couples) Act, the nation has been beset by serious storms and floods' and further explained that 'a Christian nation that abandons its faith . . . will be beset by natural disasters such as storms, disease, pestilence and war', for which reason he concluded that the recent extreme flooding was directly attributable to the then-prime minister David Cameron's support of same-sex marriage. While scientific explanations of climate change with respect to fossil fuel burning usually adopt a different tone, the tendency to attribute extreme events to human or natural ultimate causes also leads to an attribution of blame in terms of the behaviours and power structures that have led to high greenhouse gas emissions and thus caused climatic change. Both causes and effects of weather and climate change are experienced differently in different cultural contexts.

Sticking with the science for now, to approach more closely the idea that 'climate is what you expect and weather is what you get', we would have to delve further into Model Land and define climate as a distribution over a set of hypothetical parallel universes of potential weather, rather than an actual average of observations over time. If we could somehow suspend climate change at one moment in time – say, 2040 – and run the modelled Earth system for hundreds of years with 2040-weather, we might hope to make a prediction of what the probability of different kinds of weather would be in 2040: the climate. But even that is not enough, because 2040 is not in itself a steady state: the Earth system is out of balance. The polar ice sheets, for example, will be much too large in 2040 for the amount of greenhouse gas in the atmosphere, and they will take centuries or millennia to reach a new equilibrium even if fossil fuel burning ceased tomorrow. So, to estimate the climate of 2040, we need not just a set of models that have 2040 conditions, but a set of models that have 2040 conditions and that have also followed the same trajectory to get there.

This is the difference between weather and climate models: weather models incorporate only the 'fast' processes, mainly atmospheric conditions, which are changing on timescales relevant to the weather forecast. Climate models also contain representations of slower processes like ice melting, vegetation growth, the changing solar cycle and variations in atmospheric greenhouse gas concentrations. Weather models do not need to know the past, only the starting point for the forecast. Climate models generally do not need to know the starting point precisely, but they do need to know the shape of the past.

Further Research needed

Many scientific papers end with a section describing 'Further Research' that will be necessary to improve or confirm the results reported. A common refrain of policy-relevant science is essentially 'give us some more money, and we will get you a better answer'. So if, after thirty years of Further Research and exponentially larger computational resource, we have not greatly reduced the overall level of uncertainty around climate sensitivity, a key headline parameter of future climate change, what is it that is actually limiting our understanding and is there any hope for (further) Further Research to make more strides?

Two high-profile climate scientists think so. Bjorn Stevens and Tim Palmer have called for a 'CERN for climate modelling': a multinational exascale computing facility to conduct and coordinate climate model experiments using more resource than is available to an individual national modelling centre. Their argument is that the present state of play, with individual models developed in an ad hoc manner by a handful of institutions, results in models that are 'not fit for purpose' and that are not capable of adequately representing the kinds of changes policy-makers want information about in order to support urgent adaptation decisions. The success of CERN, the international particle physics laboratory hosted in Geneva, shows that a well-designed framework for such a facility can also stimulate and support high-quality research in satellite institutions. Greater co-ordination of the research programme around a single model-family could allow for more systematic exploration of some key uncertainties.

It's certainly true that the existing models have enormous limitations, and a programme to rewrite, standardise and optimise model code from scratch would no doubt uncover lots of minor errors. It could also dramatically reduce the compute time of simulations and improve the transparency and interoperability of modelling, as well as the ease of generating comparable outputs for study. The downside of creating a super-model, 'one model to rule them all', is that by restricting ourselves to a single (tweakable) model, we are no longer able to represent the kinds of uncertainties that result in differences between models.

Palmer and Stevens would probably counter this by saying that their proposed super-model would be fully stochastic and modular, with the option to (deliberately or randomly) vary different parameters, switch modules on and off, or use different physical representations for different parts of the simulation, and as such it could more or less fully capture all of the variation between existing state-of-the-art climate models and do so in a more systematic way. I am convinced by their argument that it would be better in many ways than the existing situation, but ultimately it does not break out of the paradigm that limits the methodological adequacy of the current suite of models.

To make that break, we need to bring in more independent thinking. Here are some more creative ideas, intended as thought experiments in the first instance, illustrating the range of possible models if we were not limited so much by the history of American and European atmospheric science.

First, we could tackle the dominance of atmospheric physics in climate science. OK, the (Western) idea of 'climate' refers to the atmosphere, but the atmosphere isn't the only bit of the Earth system that we care about. Could we make a global climate model based on a detailed model of the Earth's ecosystems, connected together in some way, with a very detailed carbon cycle but relatively little atmospheric dynamics? An atmospheric physicist might say it sounds ridiculous and unworkable, but the same might be said by an ecologist of the climate models we have today. As climate statisticians Jonathan Rougier and Michel Crucifix point out:

> Our lack of understanding of climate's critical ecosystems mocks the precision with which we can write down and approximate the Navier-Stokes equations [for fluid flow]. The problem is, though, that putting ecosystems into a climate simulator is a huge challenge, and progress is difficult to quantify. It introduces more uncertain parameters, and . . . it can actually make the performance of the simulator worse, until tuning is successfully completed (and there is no guarantee of success).

Ecosystem modelling has a somewhat different history. Some of the first systems to be modelled mathematically were highly nonlinear predator–prey relationships which can represent wild fluctuations in population numbers. These models have a number of individually interacting components that cannot be easily separated for the purpose of study, and they are so simplified that predictions made with them would not be particularly useful for practical ecosystem management. As such, I think the average ecosystem modeller is somewhat more humble about the possibility of representing everything than the average climate modeller.

Second, we could tackle the dominance of certain mindsets in modelling. If the same people develop the new model, it will have the same assumptions and prejudices baked into it as the old ones. How about locating the new 'CERN for climate modelling' in a developing country, and hiring some people who have *not* been trained in conventional climate science – more ecologists, more statisticians, more computer engineers, and representatives from local and national governments or businesses to keep it all real? You might object that there would be a long learning curve to get such a team up to speed, but the point is that their learning curve need not and should not recapitulate the learning curve of today's state-of-the-art modellers. The aim is to do something genuinely different.

Third, we could focus more on the outliers. Currently a lot of people put a lot of effort into trying to get all of our models to agree, and the process of doing so is certainly informative – we should continue to do this. But to pursue *only* this strategy represents a basic failure of logic. To be confident that we have the right answer, we do not need to know only that all of our current models get this answer. We also need to know that no other plausible model could get a different answer. At the same time as fitting our models to agree with observation and with each other, then, we should also be pushing them in the opposite direction. What are the most outlandish forecasts made by a model that is still plausibly consistent with other knowledge? If we find that no plausible model can, for instance, come up with a runaway greenhouse effect, then if we have made some effort to explore the space of plausible models this is good evidence that it isn't a likely outcome. This kind of research programme would be aimed at filtering the tails of the distribution (the less probable outcomes) and considering whether certain kinds of change are plausible or not. At the moment, a very wide range of possible outcomes, especially for high warming scenarios, are thought to be plausible – these include the kinds of 'tipping points' often mentioned in the more pessimistic fringe of climate literature, such as changes to the ocean circulation or rapid collapse of ice sheets. If we cannot rule these out, they become a dominant consideration for some decision-making frameworks, feeding 'doomer' scenarios of catastrophic climate change. If we could rule them out with confidence, we might increase the likelihood of agreement about climate policy; if we find that we cannot rule them out, we might take the high-end risks more seriously. But because most research is concentrated in the centre of the distribution (the most plausible physics and the most probable outcomes), we don't have a good grasp on how these outcomes would unfold, how they are related, or the timescales and preconditions that would be required.

The aim here is to find more imaginative ways of extending what climate physicist David Stainforth calls the 'envelope of possibility'. If you go and stand outside one end of a football ground and ask the supporters going in who is going to win, you will probably get a consistent answer with high agreement. Does that mean there is very high confidence? If we ask more and more supporters walking through that gate, there will be more and more supporting evidence and thus – would you agree? – a higher degree of confidence in our answer. As that sketch suggests, it doesn't necessarily tell us very much if all the available models agree with each other. What we would need to do truly to generate confidence is to assure ourselves *that no other equally plausible models could disagree*. Do the fans walking in at the other end of the stadium also give the same answer? The rather daunting challenge is to specify what the other equally plausible models are. And not just by tweaking the parameters of the model we already happen to have, because that is like asking home supporters with a wide variety of colours of socks. Stainforth worked on the groundbreaking ClimatePrediction.net, a distributed computing project that allowed members of the public to download a climate model onto their own computers and contribute their own computational time to a massive co-ordinated experiment. With this extra power, the ClimatePrediction.net model ensemble was able to vary more parameters and thus run a much wider set of 'plausible' models than the standard set of experiments run by modelling centres on supercomputers. The results showed that the model used was consistent with some quite extreme warming scenarios, stimulating other researchers to take a look at the features leading to these kinds of results, make judgements about the physical plausibility and ultimately improve the models. I hope that if Palmer and Stevens's proposal for a single international climate computing facility is ever implemented, much of the computing power will be used to push at the bounds of this envelope of possibility and to envisage alternative model structures that go beyond simply varying the parameters of one formulation.

Climate science as cognitive assemblage

Richardson's dream was of hundreds of human 'computers' performing their own calculations by slide rule and networked together into a single processing unit by a 'conductor', like an orchestra playing together. It is a fine allegory of the way that today's weather and climate models do in fact work: each point is modelled individually, taking its input from and contributing its output to neighbouring points. These days, of course, it is not human calculators but subroutines of vast numerical programmes that do the number-crunching, but it would be a mistake to think that the process is fully automated. The conductor still exists in the form of modifiable subroutines that connect model components, choosing which sections of the orchestra will play, how loudly and for how long. The experience of each member of the audience, too, is personal: it depends on whether you are close to the players or far away, on your training in musical theory and on your familiarity with the piece being played, as well as your own emotional state at the time of listening. I don't want to overstrain Richardson's musical metaphor, but I think this leads us towards an interesting conception of the relationship of – modelled – scientific knowledge with the rest of society. We do not all hear the same piece being played.

The literary critic N. Katherine Hayles has developed the concept of a 'cognitive assemblage', drawing on the work of French philosophers Gilles Deleuze and Felix Guattari. An assemblage is a system that exists by virtue of the relationships between its parts, rather than the details of those parts themselves. For instance, a political alliance, a mountain ecosystem or the supply chain of a multinational company might be described and analysed as assemblages. Hayles's description of a cognitive assemblage refers to a system that has the function of processing or interpreting information. Her example of Los Angeles's Automated Traffic Surveillance and Control system (ATSAC) distinguishes between the technological components of the assemblage (the physical state of the road network, traffic monitors, algorithms for controlling traffic signals) and the human components (individual drivers, operators making decisions, the controllers who set the overall goals of the system), but notes that they are all essential to the functioning of the system.

So if we think of climate science as a cognitive assemblage aimed at socially processing information, of what does it consist and how well is it functioning? It is engaged in sensing the physical climate and Earth system with a huge array of measuring devices, processing that information both in human brains and in weather and climate models, co-ordinating and synthesising research programmes, generating research outputs, and interfacing with national and international governance. It also generates a high-profile report by the IPCC every few years, summarising the best of our knowledge about climate and feeding it into public debate as well as into policy-making.

As with any complex system, there is a lot of feedback between different components. Research outputs influence climate policies. Social and political factors influence research funding. Extreme weather and climatic events influence political priorities. Climate policies influence greenhouse gas emissions. And so on. Although, on a purely scientific basis, the aim might be to generate as accurate a description as possible of past, present and future climate, the actions taken in pursuit of that aim are guided by different conceptions of accuracy, different ideas about which aspects of climate are most relevant and different presentations of research findings to support certain kinds of decisions. In a 2020 article, environmental geographers Duncan McLaren and Nils Markusson examine in detail the interrelations between the evolving capabilities of climate models, the technical and policy options on the table, and the political emphasis on different kinds of governance, showing how these have co-evolved to reckon with new information, perspectives and political shifts. They highlight in particular the ways that promised solutions in the past have failed to live up to their advertised potential and so 'layers of past unredeemed technological promises have become sedimented in climate pathway models'. As they say, this constant reframing and redefinition of climate targets tends to defer and delay climate action – even when the intentions of those involved are largely positive – and this undermines the possibility of meaningful responses, as a result constantly shifting the burdens of climate risks onto more vulnerable people.

From moral hazards to model hazards

There is a modellers' joke that plays on the arbitrary nature of model solutions and this kind of disconnect with the real world. It goes like this. A physicist, a chemist and an economist are shipwrecked on a desert island with nothing to eat but the ship's store of canned food. The physicist begins looking for a suitable stone to open the cans with; the chemist collects wood to make a fire; the economist thinks for a little while and then says brightly, 'Assume we have a can opener!'

This joke is not just at the expense of economists, because many complex models contain a 'can opener': a convenient way to solve the question at hand, put there by assumption, which second-guesses the problem. In itself this is not problematic: it can be a helpful way of exploring the issue at hand and considering the effect of different interventions, strategies or discoveries. The IPCC uses Integrated Assessment Models which represent the energy and climate systems and project carbon emissions to the end of this century, based on large-scale use of 'carbon dioxide removal', ways of capturing carbon from the atmosphere and storing it separately, which at the time of writing are still at the demonstration stage. That's OK – in fact, a model that shows how effective a can opener could be may stimulate us to look for the can opener or to try to develop one from the materials available, and self-evidently it would be a good thing to have such a tool. I very much hope that carbon dioxide removal at scale is possible and sufficiently cheap. One problem, which McLaren and Markusson have highlighted, is the possibility that the can opener does not exist or cannot be developed, and in that case the danger, like in the joke, is that taking the assumption too literally may discourage pursuit of necessary alternative strategies.

The carbon dioxide removal can opener is part of an answer to one scenario question that energy system modellers have been asked: if we have achieved a stabilisation and reduction of atmospheric greenhouse gas levels by the end of the century, how could it have happened? Carbon dioxide removal is certainly one answer to that question, perhaps the only answer that can be sufficiently easily encoded in energy system models and perhaps the only answer that is currently politically acceptable. But this single answer is now enshrined in models that purport to 'optimise' outcomes and offer tautological policy conclusions such as 'All pathways that limit global warming to 1.5°C with limited or no overshoot project the use of carbon dioxide removal on the order of 100–1000 $GtCO_2$ over the 21st century.' Of course they do, because the inclusion of this technology in the model at a certain (arbitrarily chosen) price point guarantees that it will be used in preference to any nominally more expensive way of reducing emissions. You could set the price point at any level in order to get exactly the use of carbon dioxide removal that you deem to be the right amount. The makers of these models have decided that the right amount is quite a lot, so that's what we get, and the results are used to argue that emissions can continue to rise because our children will be able to take the carbon out of the atmosphere later. Equally, if you could set a price for inducing behavioural change that reduced the demand for high-carbon consumer goods – maybe the Big Tech companies are working on this now as their next big product! – then that price could be set in the models and included as a carbon reduction strategy. If it were a sufficiently low price, it would be used by these models in addition to the usual suite of technological options. The point is that the entire outcome of these models hinges on a judgement that has been made that carbon dioxide removal at scale has a reasonable price and behaviour change of the wealthiest consumers at scale does not. That is a very ideologically laden statement.

And when one can opener fails to work, it is tossed aside in favour of another. As emissions continue to rise, carbon reduction has given way to carbon capture from point sources like power stations. As the atmospheric concentration of greenhouse gases has continued to rise, carbon capture at the point of emission has given way to the carbon dioxide removal that I have just described. Working our way through the chain of consequences, we are now seeing that global temperatures continue to rise, so the plausible next step in this depressing series is the replacement of carbon dioxide removal with 'geoengineering' by solar radiation management – usually meaning a large-scale project of spraying aerosols into the upper atmosphere to cool the Earth by reflecting sunlight. Again, this is set out by McLaren and Markusson:

> If politicians continue to demand scenarios that deliver 1.5C or even 2C as carbon budgets are consumed, it seems highly likely that modellers will have little choice but to include [geoengineering by solar radiation management] in the next generation of models.

All of these can-opener technologies contain a problem of 'moral hazard' – the difficulty that once there is perceived to be a quick fix, less effort is put into actually solving the root problem. Moral hazard is here reinforced by model hazard: as soon as solar radiation management geoengineering is programmed in to any of the Integrated Assessment Models, it will immediately become a go-to technology and a key pillar of the climate policy pathways that are deemed to be politically and economically feasible. Unfortunately, in purely financial terms, a very basic implementation of stratospheric aerosol geoengineering would be relatively cheap and therefore highly attractive in cost-optimising models. If the target of climate policy remains couched in the terms of global average temperature, then stratospheric aerosol geoengineering seems to me to be now an almost unavoidable consequence and its inclusion in Integrated Assessment Models will happen in parallel with the political shift to acceptability. As I write this, geoengineering is still a somewhat politically and socially unacceptable concept, and therefore it is not in the models despite being a technical possibility. But as the effects and impacts of climate change become more visible and more immediate, the realisation that greenhouse gas emissions have already dangerously geoengineered our planet may make the prospect of deliberate intervention more palatable. Geoengineering is increasingly featuring in near-future climate fiction by bestselling authors like Kim Stanley Robinson and Neal Stephenson, and Integrated Assessment Models are essentially just a mathematical version of near-future climate fiction.

If solar radiation management does become the next big thing, we will need to have thought about it carefully, ideally *before* the point that it becomes a pillar of international climate policy. Are international governance and diplomatic relations strong enough for the challenge of working out what to do when a nation suffers terrible climatic events the year after the geoengineering programme begins? If we are very lucky, the best-case scenario is that climate models will be able to show confidently that most regions will experience a net beneficial change in climate relative to a non-geoengineered world. But due to the rapidity of current change and the normal year-to-year variability, we have to expect that extreme weather events – possibly catastrophic events like the heatwave portrayed in the opening chapters of Robinson's novel *The Ministry for the Future* – would still happen *even after climate interventions have been deployed*. It would be a huge scientific and political challenge to navigate the complexity of responsibility for the effects of geoengineering, and we are not yet anywhere near a solution to the smaller and still-colossal scientific and political challenge of responsibility for the effects of greenhouse gas emissions.

So I think the can openers have to be taken seriously. Once they enter our Model Land, they become something more than just technical speculation. Just as fiction has the power to change how we think, so this mathematical version of climate fiction exerts a strong narrative pull on our political and scientific institutions.

Climate quantification and financialisation

Climate science – like most policy-relevant sciences – has become very quantitative over the last decades. The upshot is that future-oriented climate modelling results are largely of use to the most quantitatively sophisticated stakeholders: insurance and finance companies, engineering consultants, multinational companies and so on. Histograms and model sensitivity analyses for future climate are of much less use to your average subsistence farmer, concerned citizen or medical doctor. Quantitative presentation of uncertain future information is also in general of more interest and use to those who have a wide portfolio of interests over which they can experience the full probability distribution. If, for instance, you own a multinational company with 200 factories worldwide, a probability distribution for those that may be flooded in the next twenty years is something you can use to manage your risk, perhaps closing one and opening another somewhere else, and expect to come out on average better off than you would have without that information. But the owner of a single factory, unlike that of a multinational, experiences a single actual physical outcome and not the average or net effect of a distribution. Making an uncertain decision about a 1% chance of your own factory being flooded is very different from making multiple decisions about a large portfolio of factories, each of which has a 1% chance of flooding. It is even harder if you are having to make a decision about your own home, yet the negative effect of traumatic experiences and the value of emotional attachment to home and landscape rarely feature in economic analyses of climate-related loss and damage, precisely because they cannot be quantified.

Unquantifiable objects cannot be financialised. Quantifiable objects can. As quantitative modelling has become more effective, therefore, quantification of current and future climate has translated into financialisation of current and future climate. Let's look at a specific example.

In September 2014, Hurricane Odile made landfall in Mexico as a Category 3 major hurricane, sweeping across the Baja California peninsula and causing damage of US$1.25 billion. Before the hurricane season, however, a $100 million 'Catastrophe (CAT) Bond' had been issued. Effectively, this was a form of insurance: if a hurricane were to hit Mexico, a payout would be made to help with the clean-up. In the event that no hurricane occurred within the period, investors in the bond would receive a generous interest payment and have their capital returned. In order to write the agreement, the provider needed to define exactly what would count and what wouldn't. Known as a CAT-in-a-box structure, a hurricane would activate the payout if it made landfall in a specified region or 'box' during a specified time period with at least a specified intensity. The intensity of the hurricane would be measured based on the central barometric pressure according to the National Hurricane Center (NHC) official report, which is based on measurements and modelling of the region. The bond would pay out 100% to the Mexican government if the pressure was 920 millibars (mb) or below, or 50% if between 920mb and 932mb.

As it approached the coast, measurements of Odile's central pressure dipped to 923mb and estimates of the pressure on landfall were around 930mb, sparking expectations that the CAT bond would be triggered. But that's where the story starts to get interesting, because as it passed over Mexico, Hurricane Odile knocked out some official measuring instruments and very few actual data points were available. Amateur storm chaser Josh Morgerman was holed up in the Holiday Inn Express in Cabo San Lucas, very close to the point of landfall, and measured an unofficial reading of 943.1mb. The NHC took ninety-two days to produce its official report, during which time neither investors nor the Mexican government knew whether the payment would be triggered. The final report pinned the estimated landfall pressure at 941mb, based on Morgerman's reading, and finally ruled out the possibility of a payment.

This raises multiple issues. One is the sensitivity of millions of dollars of financial payments to a single observation made by a non-professional. Another is the role of the meteorologists at the NHC, who are drawn into this financial arrangement without their prior or formal agreement but whose professional judgements may make the difference between a payout occurring or not. The *Los Angeles Times* reported that James Franklin, former director of the NHC unit, had said he deliberately avoided becoming familiar with the bonds so that he could 'evaluate the meteorological data without knowing what any financial consequences might be'. But as Robert Muir Wood of catastrophe modelling company RMS observed, there is clearly potential both for well-meaning influences – perhaps a generous interpretation of data close to a threshold to ensure that an emergency response to a catastrophe in a developing nation is facilitated – and for more nefarious manipulation by those with a financial interest on one side or the other of the payout.

As with any bad disaster movie, there is a sequel. Hurricane Patricia arrived just over a year later, striking Mexico further to the south. While at sea, Patricia was a record-breaking Category 5 Pacific hurricane with the lowest central pressure yet recorded (872mb), but it weakened significantly on the approach to land and at the time of landfall the operational estimate by the NHC was a central pressure of 920mb, which would have just squeaked into the maximum payout category. The official report came down to a few sparse observations once more – and, coincidentally, Josh Morgerman was there again with a pressure reading of 937mb from a hotel a couple of miles inland. Fortunately for the Mexican government, an automatic weather station at a luxury villa nearby measured 934mb, and the NHC official report stated that the central pressure had been 932mb at landfall, triggering a 50% payout.

This raises even more issues. Patricia was in fact a somewhat less damaging hurricane than Odile had been, and yet it triggered a payout where Odile had not. At landfall, Patricia had more extreme central pressure and slightly higher wind speeds, but it struck a less populated area and had a narrower path of damage, causing only an estimated US$325 million of damage, a quarter of that of Odile. So while the financial mechanism was well suited to the investors in the bond, it perhaps proved somewhat less well suited to the issuer. In principle, the reason for investors to invest in CAT bonds is that they make a net overall long-term profit (similar to insurance providers), where the reason for issuers to provide CAT bonds is to plug short-term cash flows and allow quick responses to a catastrophe (again, very similar to insurance). If the timescale for the official report to appear is one hundred days, the short-term benefit isn't there, and the question of who is actually benefiting from the arrangement comes down to a model for how often such events are expected to occur. In that case, if there were agreement on the model, then each CAT bond would be a pure gamble for both buyer and seller rather than the kind of insurance mechanism it is usually marketed as representing.

In making the official report for Hurricane Patricia, the NHC also broke with tradition by providing a range of uncertainty for the landfall pressure, giving a value of 932mb (right on the line for a payout) with 'the uncertainty of this value likely on the order of 2–3mb'. This didn't influence the payout but it again flags a potential problem for the CAT-in-a-box mechanism: perhaps not only the source (the NHC) but also the method or model for central pressure estimation need to be more standardised to make it work, especially where data are limited and there is significant potential for disagreement. More broadly, it seems that the question of whether this kind of financialisation truly offers benefits for all parties is one that has so far been passed over in favour of making the financial mechanism work for investors. We will meet this concern again in the context of public health.

Playing DICE with the climate?

CAT bonds are formalisations of a financial approach to dealing with one kind of climatic event, but we also use models to give a much wider view of the impacts of climate change. Integrated Assessment Models of economy and climate are widely used to generate scenarios of greenhouse gas emissions from human sources, taking into account possible physical and political constraints like fossil resources, carbon taxes and new technologies.

Within Integrated Assessment Models, the costs of climate change are modelled by a 'damage function' that maps the expected global cost of impacts against the level of warming. In principle, the damage function is arrived at by adding up projected losses in different sectors, such as agricultural yields, energy requirements for heating and cooling, changes in heat-related health conditions, land loss to sea-level rise and so on. In one model (DICE2016R), the damage function is a smooth curve which estimates a total loss of

2.1% of global gross domestic product (GDP) at 3°C, and 8.5% of global income at a global temperature rise of 6°C. You might think, wait just one second – climate scientists have told us that 2°C of warming is 'dangerous' and 4°C would be catastrophic. But if the overall damage to GDP follows a smoothly increasing curve with temperature, then we are not talking about a short-term shock but something more like a delay effect that marginally slows growth and means we take a few more years to reach the same standard of living.

Those damages are experienced by a global economy which is assumed in the baseline model to continue to expand with 2% economic growth per year throughout the twenty-first century. As a reminder of scale, 2% annual compounded economic growth makes the economy more than seven times larger in a hundred years' time. Within this kind of Model Land, it is assumed that the physical climate changes *while all other things remain equal.* Beyond that, it is an essentially additive model: the economy expands, climate change has some financial costs related to global temperature which are simply subtracted from that (or in some cases added), and policies or regulation to reduce carbon emissions impose further costs via mechanisms such as increasing energy prices. Costs and benefits are measured in dollars. As a consequence of that monetisation, all losses are substitutable by consumption. In other words, if a city is flooded by rising sea levels, wildfires destroy a national park, marine ecosystems are lost or a Pacific island nation becomes uninhabitable, those losses can all be fully compensated by society simply becoming richer. Costs and benefits are also summed up globally without regard to distribution. In this Integrated Assessment Model Land, billions of dollars of climate-related losses and damages falling on people around the world are fully compensated for by the CEOs of tech giants adding equal billions of dollars to their share portfolios.

For now, let's go with the premise that the burden of allowing climate change can be quantitatively set against the costs of action to avoid it, even though they do not fall upon the same shoulders or with the same impact. If so, balancing the trade-off between the (modelled) costs of climate change and the (modelled) costs of avoiding climate change can give us a neat mathematical answer for the pathway that results in the lowest overall cost to 'society'. William Nordhaus, a recipient of the 2018 Nobel Memorial Prize in Economic Sciences for his work on climate, identified this optimal warming as 4°C in his Nobel Lecture. (Yes, you read that right: 4°C! I think he did climb down a bit to 3.5 in the publication that followed.) The Integrated Assessment Model that he has developed, called DICE (which stands for Dynamic Integrated Climate-Economy), is one of three models used by the United States Environmental Protection Agency to inform policy about climate change. Effectively, this model concludes that the scientific and political target of not exceeding 2°C change in global average temperature is an unnecessarily costly intervention whose benefits – avoided climate impacts – amount to less than the additional costs of action.

Contrast this approach with the first aim of the Paris Agreement, adopted at the United Nations Climate Change Conference in 2015, which commits signatories to a goal of 'Holding the increase in the global average temperature to well below 2°C above pre-industrial levels and pursuing efforts to limit the temperature increase to 1.5°C above pre-industrial levels.' If we take the Nordhaus model at face value (and, to be clear, I do not think we should), what it implies is that the Paris Agreement is founded on a political belief that less climate change is better *even if it costs more*. Personally, even if I were to accept DICE, I would still take the stance that the distribution of GDP matters as well as the total amount. But scientific as well as political criticisms of the DICE model have been raised.

In his 2018 Nobel Lecture, Nordhaus also stated that '[p]eople must understand the gravity of global warming'. Yet on the same day that Nordhaus was awarded his prize, the IPCC published a special report on global warming of 1.5°C, laying out the risks that we are already experiencing and the further risks of warming reaching 1.5°C and 2°C above pre-industrial levels. In effect, the argument put forward by DICE and similar models is that the first-order economic impact of losing coral reefs, mountain glaciers, sea turtles and people in poor countries is zero compared to the financial benefits of burning fossil fuels in rich countries. How might one value intangible losses such as flooded family homes, changing cultures and agricultures, loss of access to nature, the loss of species and ecosystems, or the mental health impacts of extreme events? According to the economists, this again comes down to money. Surveying the literature, economist Richard Tol wrote that

> valuation studies have consistently shown that, although people are willing to pay something to preserve or improve nature, they are not prepared to pay a large amount. Most studies put the total willingness to pay for nature conservation at substantially less than 1% of income.

If willingness to pay reflected value, we would find that oxygen is worth much more to an American than to an Ethiopian.

And we haven't yet counted any second-order economic effects such as having to decommission nuclear power stations sited next to rising seas, upgrade extreme weather protection for critical assets, teach small-scale farmers how to grow new crops with which they have no cultural familiarity, rehouse climate migrants or replace insect pollination services. In Integrated Assessment Model Land, these costs are all assumed to be zero until somebody publishes an academic study estimating them. One by one, this is happening, but each individual impact is small and is generally estimated as a marginal cost with all other things being equal, rather than in the context of a world in which all impacts are happening simultaneously. Tol and Nordhaus imply that there are absolute benefits of climate change up to about 1–1.5°C of global warming, and net benefits (set against the financial cost of mitigation policies) up to around 3.5°C. There is certainly a discrepancy here between what these economic modellers have to say about the 'optimum' climate and what many physical climate modellers have to say.

As Mike Hulme has described, concepts of 'optimal' climatic conditions have varied over time. They are invariably produced by dominant groups who cast their own original climate as being 'optimal' for human development, on the grounds that it produced such a wonderful civilisation as their own, where other climates were either conducive to disease or laziness, or simply unfit for agricultural production of their favourite foods. History, of course, is written by the winners – and so is science, mostly pursued by well-compensated upper-middle-class individuals in air-conditioned offices in temperate climates.

A more modern application of a similar approach is the statistical regression of economic productivity, measured by GDP per capita, against regional climatic variations. Needless to say, this shows that the temperate climates of Europe and North America are the most conducive to economic prosperity. Then a bit of warming isn't a problem for New York and London, and if it is a problem for Pakistan and Tanzania, well, they were not contributing much to global GDP anyway. Let us not ask ourselves (yet) whether there might be any other reasons for postcolonial lack of productivity in developing nations, as measured by distinctively Western concepts like GDP. Taking a more nuanced approach, the authors of an influential study published in *Nature* in 2015, instead of attributing absolute differences between countries to temperature variation, analyse the difference in output between cooler and warmer years in the same country. Projecting observed responses forward, the model of economists Marshall Burke, Solomon Hsiang and Edward Miguel proposes that there will be small positive economic impacts in the high latitudes (Russia, the north of North America, Northern Europe), but potentially very large negative effects in Africa and Asia, resulting in regional losses of up to 80% of GDP by 2100 for a high-emission pathway. Overall, they suggest, 'average income in the poorest 40% of countries declines 75% by 2100 relative to a world without climate change, while the richest 20% experience slight gains, since they are generally cooler'. While these figures sound more realistic and consistent with the kinds of messages coming from physical climate science, the geopolitical implications of near-total economic collapse in the poorest 40% of countries and the moral and political implications of letting this happen are – of course – outside the scope of the study.

Ceteris non paribus

The problem I have identified here is the modelling assumption that all other things can remain equal. What might we uncover if we do not automatically default to this reductionist framework?

In physical climate science, the rebounding effects are known as 'climate feedbacks'. The standard approach is to model in a forward direction, with causality running from carbon emissions, to atmospheric concentrations of greenhouse gases, to the radiative forcing experienced by the planet, to changes in climate, to changes in weather, to impacts experienced by people and ecosystems. Feedbacks happen when causal relationships are found that run the other way, connecting a later step in the chain with an earlier one. An example of a climate feedback is the melting of permafrost (a result of warmer temperatures) causing release of methane and carbon dioxide (increasing atmospheric concentrations). Another is the change in cloud patterns (change in climate) resulting in more or less sunlight being reflected back to space (change in radiative forcing). Others could be sea-level rise flooding key infrastructure, resulting in greater emissions due to reconstruction, or political change resulting in more stringent policies to reduce emissions.

The forward-chain approach suggests that uncertainty increases at each step, but the introduction of feedbacks makes everything more complicated and more uncertain. The effect of most feedbacks can only be estimated from counterfactuals (i.e., models) except in very special circumstances such as the grounding of air traffic after 9/11. And since some feedbacks are self-reinforcing, this is a problem not only for prediction but also for control. We like to think that we can both predict and control the climate. The whole of the economic literature on 'optimal' Model Land outcomes implies that by enforcing the 'correct' price on carbon we could cause emissions to halt perfectly at the point where the net cost to society is lowest. Uncertainty and self-reinforcing feedbacks make this more difficult.

The concept of a 'tipping point' describes a situation where an abrupt change happens and a system 'tips' from one state into another. Mathematically, this is identified in catastrophe theory as a system that has two stable states and that can suddenly jump between them, such as an overloaded ship which is tipped over by a gust of wind. The weight of the ship is a self-reinforcing feedback once it begins to tip, and the change in this case is irreversible without some intervention. Note, though, that despite the self-reinforcing feedback it does not necessarily imply a runaway change; the ship does not fall to infinity but is stopped once it reaches a new equilibrium in the sea. At that point it may possibly even bob back up again into its original position, but only if enough of the cargo falls off and the ship is designed in a certain way. Or it may take on water and sink further. A number of potential tipping points have been identified within the climate system, including the disintegration of ice sheets, the weakening or shutdown of the Gulf Stream and the dieback of the Amazon rainforest. These 'ships' are being overloaded by carbon emissions and other human and natural factors. Each tipping element has its own internal dynamics and some of them also contribute directly to further climatic change. When ice melts, for example, the darker land and ocean underneath absorb much more heat, and so the region warms more quickly than before. When permafrost melts, it may result in the release of methane which is a potent greenhouse gas itself. Research has looked at when possible tipping points might be triggered, but it is very difficult to model these dynamical processes, meaning that even where a tipping point can be identified it is impossible to predict with any certainty exactly when or at what level of warming any abrupt change may kick in. Models can and do make predictions, but they do not agree with each other in detail.

The level of expert judgement available to calibrate the predictions from outside Model Land is limited, because the conditions that prevailed millions of years ago are tricky to measure. We have not experienced conditions with no Amazon, no Gulf Stream or no West Antarctic Ice Sheet, so we have to develop our expert judgement based on what we see in our models.

Tipping points are also identified in a less mathematical fashion in social and cultural trends and in economics. Is there a tipping point in the social attitudes that lead to climate action, perhaps? A tipping point where renewable energy becomes cheaper than that from fossil fuels? A tipping point due to the connectivity of the internet or the capacity of artificial intelligence? Maybe we are close to all of these. The same and greater barriers prevent us from constructing accurate models of the future in these domains, so we are again left to speculate on what might or might not happen.

Without predicting in detail, we can still make useful statements. A climate change risk assessment by Chatham House produced in the lead-up to COP26 in Glasgow in 2021 took an expert elicitation approach to the cascading impacts of climate change, asking a variety of experts for their opinions in a structured way. The results offer a somewhat dark contrast with the just-a-few-per-cent-of-global-GDP pablum on the academic economists' table. The major systemic risks they identify as indirect consequences of climatic change include multiple crop failures, food crises, migration and displacement, state failure and armed conflict, none of which was mentioned even by the second academic study I described earlier, in which poorer nations lost up to 80% of their GDP. Instead of putting probabilities and timescales on the possible outcomes, the Chatham House team instead quantify the level of concern expressed by the expert panels about each: it is certainly taken as read that these are real possibilities, although I'm sure not all experts agreed on the plausibility of all outcomes. They are judgements directly made about the real world, not probabilities derived in Model Land, and they do not sit at all comfortably with the kinds of quantitative predictions offered by Integrated Assessment Models.

Reweighting the DICE

A major update to the DICE model, published in 2020 by a team from the Potsdam Institute for Climate Impact Research, shows a move towards agreement in these terms. First, it updates the physical climate module to reflect current understanding of the relationship between greenhouse gas emissions, concentrations and global temperature change – this is an 'optimistic' change, because it actually makes mitigation targets slightly easier (less costly) to achieve. Then it updates the damage function to take into account recent research on other climate damages – unsurprisingly, this one is a 'pessimistic' change and makes estimated damages larger. Next, it considers a range of expert views on the appropriate discount rate to use in the model. The discount rate represents the degree to which costs and benefits are deemed to be worth less in the future than they are at present. Part of the discount rate is essentially due to the existence of interest rates. If I have £100 now, I have the option of putting it in the bank and getting a small amount of interest, so if I expect interest rates to remain positive, then I would rather be given the sum now than wait and receive the same £100 in a year's time: it is worth more now than it will be worth later. The other part of the discount rate is an ethical judgement about the degree to which people's consumption now matters more than people's consumption in the future. If you think, for example, that people in the future will be wealthier than people now, then it may make sense to spend money now and let the future take care of itself. Or you may say that intergenerational equity means that we should value our grandchildren's quality of life equally to our own, and therefore the moral case is to set this discount rate to zero.

There are strong disagreements in the economic literature about the appropriate value of this parameter, and both extremes can cause problems for the models. Setting the overall discount rate to zero means that the far future is valued equally to the present, but since the future is potentially a very long time – after all, we have billions of years before the Earth is burned up by the dying Sun – it can become the dominant consideration. That can lead to ridiculous ideas such as diverting spending from poverty reduction today into space travel research to ensure that we can colonise other planets and therefore minimise the 'existential risk' of an asteroid colliding with the Earth and wiping out human life forever. At the other end of the spectrum, setting the discount rate to several per cent per year means that everything that happens more than a few decades from now is discounted to pretty much zero and therefore not taken into account by the model at all.

I say this to emphasise that economists are by no means monolithic in their views. The DICE update used a survey of nearly 200 working economists that showed graphically the amount of disagreement. There is no 'right' answer. Within the confines of reasonableness (which itself is somewhat subjective), the choice of a higher or lower discount rate is a subjective ethical decision and it has a first-order effect on the outputs of these models. The updated DICE with altered parameters showed that instead of Nordhaus's optimal balance at 3.5°C of global temperature increase, their economically optimal pathways using the same model were more consistent with the Paris Agreement target of limiting warming to less than 2°C.

DICE is of course not the only Integrated Assessment Model in the game, only a well-known and often-used one. Other Integrated Assessment Models make different assumptions, take different approaches, put different emphases on the relative importance of detailed modelling in different economic sectors. The IPCC's assessment reports are divided into three working groups: the first describes the physical science of climate; the second describes observed and projected impacts and potential for adaptation; and the third is about mitigation by reducing emissions through climate policies. Working Group I is largely based on the outputs of physical climate models, and Working Group III is largely based on the outputs of Integrated Assessment Models. The most recent report, published in April 2022, assesses hundreds of scenarios from models including those optimising on the basis of cost (as above) or projecting forward the implications of current policies and pledges, and concludes that the global economic benefits of limiting warming to 2°C do indeed outweigh the costs of mitigation.

Failures of imagination

The failure of most wealthy countries to address climate change ultimately represents an immense failure of imagination, and this failure of imagination has been compounded by the unimaginative ways in which science makes and interprets mathematical models.

The first failure of imagination is a lack of understanding of the scale of the problem. Climate models trot out increases in temperature that seem tiny compared to normal seasonal and even daily variability. It is therefore not surprising that people look at a forecast of 2°C (or even 4 or 6°C) change and think you would barely notice that from one day to the next, so why the big fuss? If societies and ecosystems across the world can cope with radical changes of tens of degrees in just six months between summer and winter, then adaptation to another couple of degrees should be only a small additional burden, right? If climate impacts were limited to temperature changes, this viewpoint might be just about correct. This creates a second tier of climate scepticism – not the 'it's not happening' viewpoint now overtaken by evidence, but the belief that 'it just won't be that bad'. Physical models in general struggle to predict disruptive changes to the climate because there is simply not enough previous evidence to generate confidence. Yes, models show changes in the frequency of extreme events like hurricanes and flooding, but until sufficient numbers of events have happened there is no way to calibrate the performance of the model against reality, so there is a great variation between different models and the IPCC will only assign 'low confidence' or 'medium confidence' to such projections until it is too late to do very much about it. Yes, models show a possibility that the ocean circulations could change enormously, but as we are barely able to measure present ocean currents and have only circumstantial evidence for how they have changed in the past, there is again no way to establish confidence. Unfortunately, absence of confidence is not confidence of absence. The effective result is that the possibility of substantial change is downplayed, plausible outcomes do not make it into secondary models such as economic assessments, and thus the risks being run by failing to limit climate change are starkly underestimated. And because the human consequences of climate change are modelled by different models and therefore kept in a separate category from the physical consequences, there is little focus on the kinds of cascading impacts identified by the Chatham House research team I mentioned earlier: food and water crises, displacement of peoples, armed conflict and so on. The cognitive assemblage that processes climate information is fixated on high confidence, on making perfect predictions and on physical changes rather than human impacts. So we can do fantastic science and make what are probably very good, high-confidence predictions about global average temperature, and at the same time completely fail to imagine or communicate the scale of what it could mean for humans and ecosystems in the twenty-first century.

The second failure of imagination is a failure to understand the scale of the solution required. Climate scientist and engineer Kevin Anderson has said that there are no non-radical futures any more, since we will either make radical voluntary change to decarbonise our systems or else be forced to accept the radical change that will occur due to the influence of climate and ecosystem changes on economies and geopolitics. Integrated Assessment Models of climate and economy, as we saw earlier, are incredibly conservative in terms of the kinds of future pathways that they can represent: everything must be costed, costs must be minimised and the only way to effect change is by changing the costs. As such, the only levers that can be pulled are financial mechanisms like carbon taxes which increase the cost of polluting, or speculation about how the costs of renewable energy will be reduced by innovation in engineering and new technology. In Integrated Assessment Models, as highlighted by modeller Ajay Gambhir and colleagues, there are no radical futures. Everything in these models looks like it does today, just with a bit more technology or a different price on carbon. There are no conflicts over fossil fuel resources, just as there are no groundswells of popular opinion voting in effective green leaders and changing pro-environmental behaviours. There are no shortages of the materials required to roll out renewable technologies on a vast scale;

there is no backlash against the capitalist status quo; there is no climate justice. The causes, effects and solutions to climate change, in these models, are only minor perturbations to an otherwise perfect trajectory of economic development at a rate of a few per cent a year, to infinity and beyond.

I said above that these models are just mathematical versions of near-future climate fiction. But they are mostly not very compelling stories: they do not question today's hierarchies or offer any moral reflection. That's not surprising: as we've seen in previous chapters, models tend to act conservatively and to reinforce the status quo. The whole concept of predicting the future can sometimes end up reducing the possibility of actively creating a better one. If we want the future to look different from the present, and not just a continuation of all of today's trends, then we have to construct models that are able to imagine something more.

9
Totally Under Control

> We have it totally under control. It's one person coming in from China. It's going to be just fine.
>
> President Donald Trump, 22 January 2020

Epidemic modelling, previously a fairly niche subject, has been in the public view since early 2020 and subjected to scrutiny from all sides. Although this has certainly been painful for some of the scientists involved, I think it's been particularly revealing of the ways in which public health modelling draws from and responds to its social context. As we've seen already, mathematical models inevitably contain value judgements, so the case of Covid-19 is an accessible way to start thinking about what those value judgements are and how we might improve the way that models inform decisions. The relations between modelling, policy-making and general public have been on display, at times somewhat strained and always contested.

But let's begin at a more basic level with conceptual rather than mathematical models. If you think that the recurring fevers of malaria are caused by bad air (literally, *mal aria*), the response may be to drain marshland or improve sanitation. If you believe they are caused by witchcraft, the response may be exorcisms or physical violence. If you attribute them to the effects of a mosquito-borne parasite, the response may be to spray insecticide or sleep under netting. Any of these might seem a logical response if you hold one view or bizarre and even counterproductive if you take an alternative view. Causal models of how diseases emerge or spread inform the kinds of actions that individuals and societies take in response to disease incidence. They provide a conceptual toolbox with which individuals may seek to minimise their personal risk, governments to safeguard citizens, institutions to continue to function, or scientists to corroborate or disprove hypotheses.

The causal model also offers an opportunity to make predictions: will the population be less afflicted if a certain action is taken? Will the health of a new settlement be better in one area or another? Who is likely to be most vulnerable? What other aspects of life will be affected either by an outbreak of the disease or by measures taken to prevent it? Yet the adoption of a particular model also tends to create inequalities of knowledge and power. If I have access to whatever models are deemed to be the 'right' ones, I can use that model to tell you what to do. As the mayor of a town, I can tell peasant farmers they must drain their wet fields; as a religious leader, I may exhort my followers to behave more piously; as a humanitarian following contemporary scientific advice, I may distribute mosquito nets and insecticides. More complex models generally entail greater inequality of knowledge, since not everyone has the time to invest in understanding them. They also make greater demands on existing social relationships, since they require trust in experts to maintain their legitimacy. As a result, when epidemic diseases strike, they affect populations both physically, through the direct experience of the disease, and through the way that the prevailing models mediate social and cultural responses to the disease. There are differences between the experiences of individuals (and societies) who may have different inherent biological vulnerability, and there are also differences between the experiences of individuals (and societies) who respond differently due to their assumed models of how the disease works. In order to be usefully predictive, therefore, a model of an epidemic disease must also be able to incorporate its own effects. This reflexivity puts it in the same class as economic and financial models which also directly influence the things they 'predict'. Fortunately, the usefulness of epidemic models, like economic and financial models, is not limited to their ability to predict the future successfully.

Prequels: epidemic modelling before Covid-19

In March 1918, the first cases of the H1N1 influenza virus, known as 'the Spanish flu' even though there is no evidence it originated in Spain, were recorded in Kansas. A month later cases were reported in France, Germany and the UK, and over the next two years it spread with devastating impact through a global population that had little scientific knowledge or understanding of viruses. While the 'germ theory' of disease and the model of interpersonal spread through close contact allowed many people and societies to take constructive action against outbreaks by handwashing, wearing masks, etc, only rudimentary models existed. It is estimated that the 1918 flu may have killed as many people as the 1914–18 Great War itself, and possibly a great deal more – somewhere between 1% and 6% of the global population.

Another flu virus caused a small outbreak of disease at an army training centre in Fort Dix, New Jersey, in the cold January of 1976. Thirteen young army recruits were hospitalised and one died. Samples of the virus taken from the affected soldiers were analysed by the Center for Disease Control (CDC, now the Center for Disease Control and Prevention) and found to be a H1N1 pig influenza or 'swine flu' virus, with significant similarities to the strain that had caused the disastrous 1918 pandemic. At this point, 1918 was still within living memory. Many people in the US had lost family members to the flu or knew of someone who had, with 675,000 Americans counted in the final death toll (a number only exceeded by Covid-19 in October 2021).

So there was good reason for concern. The recruits must have been infected by person-to-person transmission since they had not been in contact with pigs, and antibody testing suggested that as many as 500 of them may have been infected. This particular virus had not been seen for decades, so anyone younger than fifty could be assumed to have little or no prior immunity. Milder but large flu pandemics had occurred in 1957 and 1968, so a new and potentially more severe pandemic was absolutely expected to happen at some point. Could this be it?

In the alert that followed the discovery in New Jersey, the CDC found no other swine flu traceable to pigs: the Fort Dix outbreak was the only established instance of person-to-person transmission, and the Northern Hemisphere flu season was at an end. But the possibility of a new global pandemic emerging from these beginnings could not be discounted. Scientists nervous about the prospect of the winter season to come were unable to rule out the worst-case scenarios. Manufacturers of vaccines would need the whole summer to produce enough doses to vaccinate most of the population, so some kind of decision had to be taken very quickly. Politicians could not stand the possibility that something very bad might happen *after* they had been warned about it sufficiently in advance to have had the opportunity to do something. As one observer put it, with those people in that situation it was 'politically impossible to say no'. So the machine rolled into action. President Ford ordered a national immunisation programme; manufacturers began producing doses; public health officials began planning for mass immunisations . . . and then the insurers of vaccine companies decided that they would not take the liability for the possible outcomes of immunising everyone in the country, and the whole plan began to unravel.

With a completely unrelated outbreak of Legionnaires' disease in Philadelphia that for a few weeks looked as if it might have been swine flu, enough political momentum was generated to indemnify the vaccine manufacturers, and 40 million Americans were vaccinated, including President Ford and his family. But with no reappearance of swine flu and mounting concern about side effects of the vaccine (some real, some coincidental), the project was halted. Ford was defeated by Jimmy Carter in the November presidential elections, there were various changes of personnel at the CDC and the remaining doses of vaccine never made it into circulation. Nor did the swine flu.

This story is relevant as a contrast to the way models are used today. In fact, it's striking that quantified risks and any form of future prediction hardly featured in 1976. Public decision-making did not use significant computational aids to predict the course of the outbreak, only plausible scenarios.

Surveillance of viral influenza circulation in both humans and animals has greatly improved in the decades since then, so when an outbreak of H1N1 swine flu was detected in early 2009 in Mexico, public health systems across the world were not taken by surprise. In 2008, the UK's National Risk Register even included pandemic influenza as the top-rated risk to the country (very high impact, moderately high likelihood) and recommended the stockpiling of antivirals (Tamiflu) in preparation. So the UK response followed a playbook that had been rehearsed already, although it seems that the expectation was of a slightly different H5N1 bird flu rather than another H1N1 swine flu.

And, in contrast to the epidemics of the previous century, mathematical modellers swung into action. A team from Imperial College London who had previously provided advice for the foot-and-mouth disease outbreak in cattle in the UK in 2001 were quickly called upon to produce models to forecast the progress of swine flu in the human population. Feeding into official committees including the Scientific Advisory Group for Emergencies (SAGE) and its subcommittees, these models became effectively the primary source of planning assumptions about the severity and spread of the disease, and also about potential impacts.

Dame Deirdre Hine wrote a retrospective official review of the UK's response to the 2009 swine flu pandemic, in which she notes the importance and influence of modelling in informing action by forecasting possibilities, developing planning assumptions for operational decisions and suggesting 'reasonable worst-case scenarios'. For the most part Hine's assessment of the contribution of modelling is very positive, but she notes that in the early stages of the pandemic there simply wasn't enough data to calibrate the models to say anything with confidence. Perhaps more importantly, she concludes that the emphasis on modelling 'reduced the opportunity for a full contribution by other disciplines' such as clinical epidemiology and behavioural science, because the implementation and discussion of theoretical models crowded out the limited time and attention of policy-makers to the detriment of other sources of information.

Like the US response to the putative outbreak in 1976, the 2009 pandemic response in the UK has been criticised as an overreaction based on alarmist scientists 'crying wolf' or exaggerating the threat, with the government spending millions on antiviral drugs and vaccines to have the capacity to treat and vaccinate a very large number of people. In the event, 457 deaths were recorded in the UK. Despite the swine flu epidemic going global, it turned out to be a strain of a similar severity to normal seasonal flu, the main difference being that it affected a larger number of younger people. That information, however, is only available with the benefit of hindsight and, as Hine describes, the uncertainty at the time meant that it was again politically impossible to say no:

> Ministers in particular told [Hine] that, in the absence of greater clarity about the nature of the virus, its potential to mutate, and the likely impact on different groups, . . . the only decision they felt comfortable making was to purchase enough vaccine to cover 100% of the population.

The possibility that something very bad could have happened and in the event be shown to have been avoidable was worse than the prospect of being perceived to have overreacted. The political consequences of the worst-case scenario outweighed other considerations.

Prophets of doom

It has been said that modellers always focus on the worst-case scenario. Whether it is in relation to climate change or potential pandemics, in some circles they seem to have been cast as modern-day Cassandras prophesying doom at every opportunity. Here I think it is important to distinguish two uses of modelling: to inform decision-making and to inform scientific understanding.

As a tool to inform decision-making in the manner of the SAGE modellers, a reasonable worst-case scenario is relevant, as are all the other possibilities. In this context the decision to purchase many doses of vaccine or antiviral treatment can be seen as an insurance policy against the small but plausible risk of a high impact if it were to come about, rather than a declaration that the worst-case scenario is the expected outcome. Of course, there is a necessary debate about the price of the insurance relative to the likelihood and impact of the outcomes, but this is a political debate which is (or should be) had outside the sphere of the modelling itself; outside Model Land. The asymmetry of the situation must also be taken into account. For many large-scale events there are compounding impacts which mean that a doubly large event (twice as much climatic change or twice as many people affected by a pandemic) incurs *more* than double the costs. If a small number of the population are prevented from carrying out their work, for instance, there may be no interruption to services, but as the number increases there will come a point where an inability to substitute staff leads to long delays that carry over into other services and incur high associated costs. A small amount of climatic change may be dealt with by incremental changes; a larger amount requires full-scale redesign of critical infrastructure. Even if you are taking a completely symmetrical approach to the uncertainty by computing a simple expected cost (sum of probability times impact), the higher costs associated with the larger events mean that the worst-case scenario can become a dominant factor in the overall decision.

As a tool to inform scientific understanding, the worst-case scenario is also relevant, as are all the other possibilities. The worst-case scenario is probably also the one that pushes your model furthest away from the conditions in which it is calibrated and understood, and so examining the performance of the model on the worst-case scenario may identify key weaknesses or inconsistencies or the breakdown of assumptions. For the same reasons, it is also likely to have the highest associated range of uncertainty. If models are only tested on conservative possibilities, we could come away with unwarranted overconfidence in their capabilities.

The problem with worst-case scenarios isn't the science, it's the way they are reported and perceived. By definition they give interesting, perhaps scary, results that draw the eye and the mind away from less dramatic outcomes. The mistake of treating a worst-case scenario as a likely outcome is not a scientific mistake but a communication problem. At the end of September 2020, with cases rising and no vaccine yet available, the UK government's scientific advisors held a press conference in which they showed a possible future scenario (taking pains to say that it was not a prediction) of infections rising on a smooth exponential curve and translating into thousands of deaths per day over the winter. This was widely slated by the media as being alarmist, but proved to be sadly accurate as the UK faced an extremely challenging winter period, losing more citizens to Covid than in the first wave and having to enter a further period of lockdown. The next autumn, with a mostly vaccinated UK population but concerns about lack of immunity to the new 'Omicron' variant, similar concerns were raised. Fortunately, this time the worst-case scenario did not emerge: a huge spike in infections was accompanied by a smaller rise in hospitalisations and mortality.

A further problem is the reflexivity already mentioned: if models are used to change behaviours, they change the outcomes. Neither modelling to inform decision-making nor modelling to inform scientific understanding necessarily requires modelling to predict the future accurately. As SAGE scientist Neil Ferguson was quoted as saying in 2020, 'we're building simplified representations of reality. Models are not crystal balls.' Referring to the role of experts during the 2009 swine flu pandemic, Hine used a similar image: 'modellers are not court astrologers'.

It's interesting that this kind of metaphor turns up here, because what everyone would like is for the model to be a crystal ball into which we can look to divine the future with reasonable accuracy – and it's clear on scientific grounds that in the case of pandemic forecasting we don't have this. American conservative political commentator Rush Limbaugh encapsulated concerns about worst-case scenarios early in the pandemic (10 April 2020):

> Somebody who predicts the worst cannot lose because no matter if he's wrong, he can credit his prediction for scaring people into doing the right thing to make his prediction wrong. So when the prediction's wrong it's even good news. That's what I mean by doomsayers can never lose.

Yet where they are christened Cassandras or accused of crying wolf, it's only fair to point out that the moral of both of these myths suggests an opposite conclusion: Cassandra turned out to be right, and the wolf did eventually come and eat the sheep.

To return to a previous theme, the question is not whether the prediction was 'correct', but whether it aided our thinking and helped suitable action to be decided. Would our decisions have been better or worse if no models had been available? If we had had no recourse to a model that could attempt to predict the future evolution of the pandemic, would we have made more or less effective decisions about mitigation? This counterfactual is difficult to imagine: presumably we would still be able to think through the mechanics of viral transmission and the effect of social distancing even without writing it down as a set of equations. Widespread availability of mathematical modelling has hardly given clear answers about strategy: the huge variety of political approaches to Covid-19 is testament to that.

So while the ability to follow through our assumptions into quantitative plausible futures – including worst-case scenarios – is genuinely useful, I also think we have to take seriously Limbaugh's challenge that 'The modelers can't be wrong . . . They have no weight on their shoulders. They're never gonna be held ultimately responsible for whatever happens here.' He's right: modellers are not going to be held responsible for political decisions, nor should they be. But there is a responsibility for misunderstandings or failures to communicate the limitations of models and the pitfalls inherent in using them to inform public policy-making. If modelling is to be taken seriously as an input to decision-making, we need to be clearer on this front, and part of that is acknowledging the social element of modelling rather than taking it to be a simple prediction tool that can be either right or wrong: again, 'an engine, not a camera'.

What's not in the models

Many people became armchair epidemiologists in the spring of 2020. If you didn't, let's say you've now decided to make a model of the spread of a new virus in some population. Where do you start? Perhaps, taking a lead from published epidemiological modelling, you simplify the population into Susceptible, Infected and Recovered individuals (an S-I-R model), with one index case Infected individual, a large number of others who are initially Susceptible and transmission probabilities between each state. Perhaps you also include categories for individuals who have been exposed to the virus (E) in order to allow for an incubation period, and want to count the number of deaths (D). You run your model. OK – well, this looks bad, if there is any outbreak at all either it disappears quickly or we have an exponential growth and most of the population is going to get the disease in pretty short order. What can we do about it? Within this model, we can frame interventions as:

- reducing the probability of an individual being exposed (behavioural changes to reduce the connectivity of the population);
- reducing the probability of an exposed individual becoming infected (vaccination, use of face coverings, general health and dietary improvement); and
- increasing the probability of an infected individual recovering (vaccination, better medical treatment).

Let's take stock of what's been done there. We've created a model that looks somewhat like a population, but we have flattened along almost all dimensions and kept only those that are relevant to the physical course of a disease in a single individual, who progresses from S to E to I, and then hopefully to R rather than D. In extending the model from individual to population, we've assumed that all individuals are the same and that the probabilities of moving between each state are the same for everyone.

This is the basis of the epidemiological models that informed the advice given by SAGE to the UK government in the early days of the Covid-19 pandemic. In practice they are much more complicated, dividing the UK into blocks of population density, defining where individuals live, how far they travel for work and how often they come into contact with other people of different ages. A lot of information about people and their behaviour goes into the model, as well as the characteristics of the virus itself.

The actual progress of the Covid-19 pandemic, however, proved rather different from that predicted by the models, in which the spread and transmission of the virus closely paralleled population density. For example, in the UK – where I live – it may have been spreading more or less unchecked until the middle of March 2020, but during the first lockdown some areas were harder-hit than others, notably poorer regions in the North of England and the Midlands. In India, when the deadly second Covid wave spread in spring 2021, reports emerged of migrant workers making arduous journeys across the country to return home after their workplaces and public transport were closed due to lockdowns. At the end of the second wave, Maharashtra, the region with the largest migrant worker population, also had the highest recorded excess death toll.

There's a lot more that could be said about the ways in which social and political inequalities affect data availability and interpretation, but let's leave those to other discussions without downplaying their huge importance. Here, we are talking about the mathematical models based upon the data. The decision to incorporate some variables and not others is ultimately a sociopolitical choice as well as a scientific choice, even if the model is made before there is any political interest in it. The decision to compartmentalise in certain ways above others reflects the priorities and assumptions of the person or group making a model, or sometimes those of the people who have collected the data that become the only possible input to a model. In making this model, the modeller imagines their own experience of travelling from one place to another, coming into contact with other people in the household, school and workplace. Is it a coincidence that Ferguson's highly cited 2006 paper in the journal *Nature* outlining a model for infection spread identifies care homes and prisons as having a particular lack of data and that care homes were particularly hard-hit in the first wave of the pandemic in the UK and prisons particularly hard-hit in the US? A contributing factor to those data being unavailable is that residential institutions such as care homes and prisons are more likely to be outside the lived experience of your average mathematical modeller. This again speaks to the importance of diversity in the modelling community, not just for equality and diversity as a desirable end in itself, but as a means actually to improve the relevance of model inputs, outputs and predictive ability. Diversity of age, social class and background is as important as that of gender and race. If mathematical models tend to homogenise, is that because homogeneity is the ground truth or is it because homogeneity is the lived experience of those constructing the model and the easiest thing to represent mathematically?

What else can we say about the creators of these models and their blind spots? If modellers, because of who they are, can more easily imagine themselves as elderly people susceptible to viral disease than as single mothers working in low-paid insecure retail jobs with no outdoor space for their toddlers to play at home, the implicit values of the model will prioritise the former over the latter. If modellers, because of who they are, are able to work from home on a laptop while other people maintain the supply chains that bring them their lunchtime sandwich, they simply may not have access to all of the possible harms of the kinds of actions that are being proposed. This is not a conspiracy theory. Instead, this is what data scientists Catherine D'Ignazio and Lauren Klein term 'privilege hazard'. It reflects no malice aforethought on the part of the modellers, who I am sure were doing their level best at a time of national crisis. Nor does it necessarily mean that the wrong information was given: after all, the modellers provided what the politicians said they wanted, and in a democracy it is the job of the political representative (not the scientist) to combine the available information with the values and interests of citizens to come to a decision. The lack of diversity in our political class, and the demographic similarities between political and scientific elites, are certainly also problematic for the same reasons, and contributed to the lack of challenge or pushback offered to these models in high-level briefing rooms. As Chimamanda Ngozi Adichie said, 'Power is the ability not just to tell the story of another person, but to make it the definitive story of that person.' Where models are both telling the story of other people and directly influencing their future stories, there is a responsibility to seek enough information to be able to tell them from multiple perspectives rather than a single one.

With models in some sense acting as a projection of the internal judgements of the modeller, it is interesting that initial assumptions were that not all of the population would comply with lockdown restrictions; in Ferguson's 2006 influenza model, the assumption was only 50% compliance with voluntary quarantine. As far as I can tell, this was the initial assumption used for the UK's Covid-19 decision-making, which also chimed with the supposed advice given by behavioural scientist members of SAGE that there would be 'lockdown fatigue', i.e., that restrictions would only be accepted for a short time, after which citizens would simply give up on the rules. Lockdown was therefore delayed by longer than it might have otherwise been. Later studies suggest that enforcing a lockdown in the UK just one week earlier would have saved thousands of lives and actually reduced the duration of the lockdown, but the model on the table at that point held that full infection was inevitable, restrictions would not be enforceable for longer than a few weeks, and the main goal was to keep the unavoidable peak lower and flatter to try to avoid collapse of essential health services. In the event, as SAGE member Stephen Reicher has described, even less-vulnerable members of the British public turned out be much more willing to comply with restrictions for the benefit of others than was expected. Indeed, when a series of high-ranking British officials were discovered to have broken the lockdown regulations themselves, the public outcry was enormous. High-profile stories included Neil Ferguson resigning from SAGE for having a lover visit him, a senior advisor to the Scottish government resigning for visiting a second home, a senior advisor to the British prime minister heaped with ignominy but refusing to resign for having travelled several hundred miles to a second home while actually infected with Covid, and gradual revelations of multiple illegal social gatherings being held in Downing Street itself during the periods of lockdown. Do the low compliance rates initially expected by modellers reflect their poor judgement of 'the public', or is it instead a projection of how likely they feel they and their friends would be to comply with requirements in full? Both of these conclusions are disappointing. They also point to an urgent need to incorporate greater public input into overall political value judgements and into the modelling process. This is a gigantic failure of the model, at least as large as the uncertainty about basic characteristics of the virus itself, which were studied in detail.

Perhaps we should consider what else is not in these models, either for want of data or lack of interest. Where are the data on mental health outcomes for young people? Although it's a key impact that has been often cited by both sides in the debate about reopening schools, at the time of writing (and well after several rounds of reopening and reclosing schools in many places) it's still remarkably hard to find data about how the various impacts of the pandemic have resulted in changing patterns of mental health outcomes for young people or anyone else. Similarly, the homogenisation of impacts into overall mortality rates and lost GDP hides massive distributional changes that cannot be seen unless we are able to disaggregate the data by sex, race, affluence and so on. Eugene Richardson and colleagues describe the 'symbolic violence of R_0' and other model variables which, due to their mathematically attractive simplicity, obscure differences, prevent the identification and challenging of systemic disparities, and sustain relations of oppression. The same symbolic violence is at work in climate policies that focus on global mean temperature, compressing the experiences and tragedies of billions into a single net scalar variable. Any trade-off of these variables against 'the costs of action' represents a trade-off of lives and livelihoods that are not randomly chosen but always disproportionately come from the poorest and most vulnerable communities.

Unequal impacts of financialisation using models

When they inform public policy, the actual use of models can be somewhat opaque, but in other cases they are used much more directly to decide real outcomes, and here it is just as important to consider their limitations.

In 2020, like many other events, the Wimbledon tennis tournament was cancelled. But despite the loss of the major event in its sporting calendar, the All England Lawn Tennis Club posted an operating profit of £40 million. It was well prepared: from 2003 to 2017, after the first SARS pandemic occurred, it paid an annual premium of about £1.5 million to insure itself against cancellation of the tournament due to pandemic, an investment that paid off handsomely with a payout of over £170 million.

Like other insurance policies, pandemic insurance can only be offered by a company as a significant business line if the risk is well understood – and to understand risk well enough to quantify it requires some kind of model. Now that an event has happened, pandemic insurance is presumably much more popular as a business continuity arrangement. Equally, the probabilities have probably been revised up and the insurance will be more expensive and policies carefully worded.

Like the CAT bonds discussed earlier, pandemic insurance is a way of financialising crisis using mathematical models. Since insurance companies exist for the purpose of making money, this can only happen when they have a good enough model to be confident about the actual risk and add a margin on top of that to maintain the business.

Medical anthropologists and public health experts have written about the World Bank's Pandemic Emergency Financing Facility (PEF), a financial mechanism designed to secure money from international investors and release it to support timely public health interventions in the event of an infectious disease outbreak. The PEF was conceived as a response to the difficulties experienced in funding the response to the Ebola epidemic of 2013–16 which resulted in over 11,000 confirmed deaths (and likely many more unreported), mainly in Liberia, Sierra Leone and Guinea. The financing of the facility combines a simple cash fund with an insurance-like 'pandemic bond' issued to investors, who receive a certain return during periods when there is no pandemic but forfeit all of their capital in the event that a pandemic occurs. Despite being described in some circles as an insurance arrangement, the pandemic bond is rather different from the true insurance bought by the All England Lawn Tennis Club to cover the Wimbledon tournament. Instead of being based on the actual loss suffered, the PEF is a parametric bond much more like the CAT bonds that failed to pay out after Category 3 Hurricane Odile hit Mexico in 2014. The capital would only be released for pandemic mitigation measures after a set of conditions had been triggered: one of a small number of specified viruses, confirmed in an outbreak of a particular size, with fast growth of the outbreak and spread to multiple places.

The PEF was set up with good intentions. It mobilised cash that would not otherwise have been available, multiplying the money available from donors by using their input to cover interest on a larger potential total. In principle, it would release money early enough to take action to mitigate the threat, in a transparent and accountable way laid down in advance. It would incentivise countries to improve monitoring and early detection of outbreaks in order to benefit from the available cash. World Bank president Jim Yong Kim championed the programme as a way to 'insure the poor', and the first three-year bond was fully subscribed when issued in June 2017.

Concerns were raised just a year later when the bond failed to be triggered despite the return of the Ebola virus to the Democratic Republic of Congo (DRC) in the second-largest outbreak on record. A grant was made from the cash fund, administered separately, but bond-holders retained their capital and continued to receive interest paid by the donors – international aid from the governments of Australia, Germany and Japan. Susan Erikson, an anthropologist who has worked both with international financial agencies and with communities affected by Ebola in West Africa, notes that the primary driving force of the instrument was always financial, and that officials from the intended 'beneficiary' countries were not consulted in its making. Having done fieldwork in Sierra Leone, Erikson is also familiar with the data collection processes that are largely ignored by the financial mechanism but that are essential to its operation. She describes the difficult conditions in which enumerators may have to walk many miles on dangerous roads to count and report the number of cases in a village, and the possibilities for error either accidental or – given the millions of dollars at stake on both sides of this equation – deliberate. And we can see again the parallels with the CAT bonds and the necessity of measuring reliably the data upon which so much depends. The lack of care with regard to the data, after the investment of so much time and effort into the mathematical model, is indicative of the balance of power in the arrangement. Erikson writes:

> A deep reckoning with the pandemic bond forces an essential confrontation: for some people, the stakes are about losing or gaining money; for others, the risks are death or long-term disability. The numeric triggers obscure that difference, thereby creating another risk: losing sight of which risks to prioritize.

In 2019, academic economist Larry Summers, a previous chief economist of the World Bank, described the pandemic bond scheme as 'financial goofiness in support of a worthy cause'. Failing to pay out in the event of an Ebola outbreak in the DRC, it had instead paid annual interest to investors of between 6.9% and 11.1% above the LIBOR interbank lending rate (about 2% over the 2017–19 period), a return considerably better than that generated by so-called 'junk bonds'. The bond class paying 6.9% was capped at 16.7% loss while the 11.1% returns were only offered to investors willing to shoulder a possible loss of 100% of the initial investment in the event of a pandemic.

So, if it didn't release any cash for Ebola, you are probably wondering what happened to the pandemic bond when Covid-19 came along, since a three-year bond issued in June 2017 reached maturity in 2020. The coronavirus family of diseases were – luckily! – specifically covered under the terms. The bond rapidly lost value in the early months of 2020 since investors could see the likelihood of the conditions being met, and indeed in the middle of April 2020 the mechanism was triggered. This released $195.84 million, which the World Bank split between sixty-four lower-income countries that at that point had reported cases of Covid-19. Why did it take so long to respond? This was due to the trigger conditions, which required a delay of at least twelve weeks beyond the declaration of an outbreak. The start date was defined as 31 December 2019, so the end of March would have been the earliest opportunity. The second condition required a case growth rate above zero, and therefore the collation of another two full weeks of data; data that themselves are often available only in retrospect as cases are tested, laboratory-confirmed, reported to national systems and then declared to international data monitors. So, while it did trigger as soon as it plausibly could have done for this pandemic, it was dramatically too late to be of meaningful use. The money released did support some efforts but, to put it in context, by that time the World Bank itself had deployed $1.9 billion of emergency support to lower-income countries and was preparing to launch a further economic programme of up to $160 billion. Overall, although the exact figures don't seem to be available anywhere, it looks like World Bank donors contributed $72.5 million and investors contributed $420 million up front, with $196 million being paid out to support the Covid-19 response, $229 million returned to investors and something of the order of $60–70 million being paid to investors as interest. Even those investors who lost 100% of their capital would have made back around 25% in the two interest-paying years.

So, would you rerun it? At the time of writing, the World Bank says it has no plans to consider a PEF 2.0. But it's clear that the financialisation of these forms of human risk is a trend that is set to continue. Some have suggested that the PEF insurance mechanism was 'designed to fail' due to the flaws in the mechanism and the difficulty of reaching the payout criteria, but Erikson notes that 'everyone I interviewed who had worked to create the PEF – to the person – believed it was a moral good of a high order'. It is easy with hindsight to criticise the programme; it is another thing to design a mechanism that can work effectively for all parties, given the deep uncertainties about what the future may hold and the radical insufficiency of models in quantifying risk. Given the exponential behaviour of a pandemic, we are left either to act too late based on data or with deep uncertainty based on a model.

Models as boundary objects

Mathematical models such as those used to predict the spread of Covid-19 infection act as boundary objects, which facilitate the translation of information between one community and another. Epidemiologists and public health experts create a model that then focuses public attention on mortality rates and the pressure on emergency care, shapes the kind of policies and interventions that are considered, forms a visual and authoritative-seeming basis for communication with the public, and helps to construct narratives of future scenarios between which politicians can decide. Tim Rhodes and Kari Lancaster, sociologists of public health, have described the way 'models create a space, a working relationship, in which dialogue around an innovation or speculation is made possible'. Given that the model takes on so much power to shape these relationships and the content and outcome of dialogues, it is particularly important that it should be subject to democratic criticism. The only way to achieve that, if we do not expect everyone to become expert in mathematical modelling, is to improve the representativeness of modellers.

Acting at the boundary between science and policy, the model is a framework for non-expert policy-makers to become more acquainted with the relationships between possible interventions (decisions) and some of their possible outcomes by observing the results of different kinds of scenarios. It is also a framework for the policy-makers to request further information along the lines of 'what would happen if we did this instead of that?' and for the scientists to respond either by altering the model or by describing its limitations.

At the boundary between policy and public, the model is a persuasive tool, simultaneously stick and carrot: here, on the one hand, are the terrible things that may happen if we (or maybe you) do not act in a certain way; and here, on the other hand, are the better things that may happen if we (or maybe you) do act. The outputs of the model are visual supports for the implementation of preferred policies. As computer models become more and more central to decision-making, it is notable that the visualisation of model output is increasingly seen as a priority, albeit one that scientists typically say they are not very good at. Visualisation usually means tracking single model runs over time, which gives the impression that the main scientific use of the model is for prediction – an impression that, as we have seen in the previous chapters, is quite wrong. The use of models is much more nuanced and complex than simple prediction. When the accountability of models is questioned with reference to outcomes that do not precisely line up with modelled scenarios, the visualisation of uncertainty becomes a high priority. We can see that this has developed for epidemiological models when they reach the public gaze, just as it has done for climate models. But it still tends to be uncertainty around predicted development over time rather than other kinds of uncertainty, so that the other uses of the model are not visible to the non-expert.

All this is a response to feedback from the increasingly well-informed public realm seeking to understand political decisions in order to be able to assent or critique. In the past, perhaps there has not been such a direct connection between science and public due to filtering through elite institutions, academic journals and technical jargon. Now we have science communication, social media and an expectation of visibility of at least some of these figures. Writing on Twitter, Neil Ferguson described his model as 'thousands of lines of undocumented C [code]', but it has since been tidied up, checked through in detail, extended and published for open-source scrutiny and public use as CovidSim, available through a web-browser interface. How many other thousands of lines of undocumented code are on academic computers, informing smaller-scale policy decisions in an opaque and unaccountable way? I do not know, but if the Covid experience encourages some of those modellers to tidy up their domain and make their models publicly accessible, it might be a good thing.

The viral spread of models

If models develop in the primordial soup of the jargon-filled academic environment, in which metaphors collide, theories are tested and discarded, and models are developed and abandoned, how is it that some of them manage to emerge and travel so effectively into the everyday environment of our social and political lives?

The qualities that make certain kinds of scientific models well adapted for this transition from the academic sphere also give them purchase in two other domains: policy and public. The policy domain likes models that have a clear and limited set of inputs that are relevant to policy levers. It also likes models that have a clear and limited set of outputs that are directly policy-relevant, have an obvious direction of good and bad, and do not conflict with each other. Beyond that, it likes models that are backed by elite academic institutions and presented by polished communicators who can give a three-minute non-technical elevator pitch, and who recognise that they are only one of many inputs to a political decision. The public domain also likes models with these characteristics, but with greater emphasis on visualisation and communication. In particular, graphical representations that can be taken in at a glance, plotting tangible variables with intuitive colour schemes, help models travel across the scientific boundary and take a place in public discourse in their own right.

Covid models did that in spring 2020 via the idea of 'flattening the curve'. Widely shared diagrams showed infection rates plotted over time, conveying the increasing rate of spread, the need to reduce it and (in some versions) a critical threshold at which health services would be overwhelmed. This helped to cement a conviction narrative for action in the minds of both governments and the public, encouraging compliance with the imposed restrictions.

When models go viral, they bring with them patterns of thinking. The simplifications embodied in the most popular models become natural defaults. Alternative models look clunky, overcomplicated or not sufficiently rigorous to warrant action. Policy-makers begin to think of a particular type of model as a 'gold standard'. The solidification of one set of assumptions becomes something that is harder to escape from.

Polarisation occurs when one group accepts the assumptions of a certain set of models and another rejects them. Consider the difficulty of conversations based on models of vaccine effectiveness, where there is a significant group of people that effectively does not accept the small possibility of vaccine injury and another that considers it to be more important ethically than a greater risk of injury due to a disease itself. These disagreements also clearly project political alignments, especially in already polarised societies. The purpose of the model may be to help me decide what choice to make personally, or it may be to decide top-down public health strategies leading to pressure on individual choice. Is it possible to find a model that can be a space for dialogue between these two constituencies? I think it is – but that model would need to include explicit representation of vaccine injury in addition to disease implications, would need to facilitate explicit discussion of the relative value weighting of different outcomes, would need to distinguish public health outcomes and individual outcomes, would need to be codeveloped with representatives from both groups who can ensure that their concerns (whether or not those concerns are shared by the other group) are addressed, and would need to be transparently free from political or institutional pressure. Then we might have a starting point for reasonable discussion about the conflicting ethical judgements that result in different preferred actions. Note that none of this necessarily involves collecting more data, buying a larger computer or using whizz-bang statistical methods. It's about representation, taking other views seriously and building trust. Maybe one day that could go viral.

Models always contain values

If there is one takeaway here, it is that looking at models in public health can throw into particularly sharp relief the political and institutional pressures that act on scientific research, shaping the kinds of models that are constructed and the way that they enter decision-making and public conversation. When models are used to persuade as well as to inform, care is needed to ensure that they do not cross the line into manipulation or coercion. Models in public health refer clearly to benefits and harms of different courses of action, but no model can 'decide' what to do until the relative weightings of different kinds of benefits and harms are specified. Science cannot tell us how to value things; doing so involves purely human moral judgements. As such, the idea of 'following the science' is meaningless.

When the citizens of many countries were told in early 2020 that they must immediately comply with a pandemic lockdown period and business closures, the primary call to action was a moral one: we must save lives; we must protect the vulnerable; we must avoid intolerable pressure on health services; and so on. And the call was answered magnificently by citizens who made a great deal of personal sacrifice in support of the greater good. Yet the moral case made by the models was partial and biased; by construction, it took more account of harms to some groups of people than others.

No model can be perfect. And even if we were able to list exhaustively all possible harms and model them effectively, we cannot necessarily achieve any consensus about values. People simply disagree. That is unavoidable. But in order to come to a decision about action, we must decide on some set of values, whether that is by default, diktat or democracy. In a democracy that aims to make decisions in the best interests of citizens, those values should be transparent, and the voices and interests of all citizens need to be heard and weighed together on an equal basis. These political processes lie outside Model Land, but they are at least as important as the mathematics; generally much more important. Only then can we maintain trust in both scientific models and political decision-making processes more generally.

10

Escaping from Model Land

[P]rogress means getting nearer to the place you want to be.

C. S. Lewis, *Mere Christianity* (1952)

What does progress actually look like in this context, if the place we want to be is one that usefully informs decisions in the real world? Where are the exits from Model Land and how do we navigate them gracefully to make use of the immense gains in scientific understanding and insight that models can offer, without running the risk of taking our models too literally?

Model results are dependent on trust

In December 1999, a catastrophic rainstorm affected Vargas (now renamed La Guaira) State in Venezuela, resulting in mudslides, flash flooding, the destruction of entire towns and immense loss of life. After the event, in an attempt to understand what had gone wrong, mathematician Stuart Coles looked at the frequency of extreme rainfall events in Venezuela. He found that, according to a model fitted to previous rainfall measurements alone, the chance of an event like the Vargas tragedy was one in 17.6 million years – so unlikely as to be deemed essentially impossible. But it happened. Rainfall on a single day was three times greater than any previously observed, and of course the only possible conclusion is that the model was wrong. Having falsified the old model, the statisticians then returned to the data and fitted another model according to which the observations were at least plausible, even if still a rare event.

A 1-in-17.6-million-year event is still somewhat more plausible than the 'twenty-five standard deviation moves, several days in a row' that were noted by David Viniar of Goldman Sachs after the financial crisis of 2008. This remark is often quoted with a bit of a snigger because almost any useful model would be falsified by these observations. When something happens that is vanishingly unlikely, usually you do not say, 'My goodness! How terribly unlucky we were.' Instead, you have to say, 'Ah! My model was wrong.'

Unfortunately, things are rarely quite so clear-cut. If you observe an event that is not vanishingly unlikely but just somewhat unlikely, how do you know what to do? Was the unprecedented British Columbia heatwave of 2021 a 1-in-a-1,000-year event, or was it instead a sign that the weather and climate models used to understand the range of plausible events were inadequate? Probably a bit of both. Say you are an insurer, and you've paid an expert consultant for a flood loss model that gives you a threshold for a 1-in-200-year event. You write insurance contracts according to that assumption, and the following year a large flood severely affects southern Germany, where you have a lot of business. What you do next is going to depend on the trust that you have in that model. If you have a strong prior belief that the model is correct, you write off the loss as a 'rare event' and continue to insure according to those assumptions. If you only have a weak trust in the model, you fire the consultant and get your information from someone else next year (and maybe stay away from flooding hazards).

Try the same thing a different way: I flip a coin and ask you to tell me the probability of it being heads or tails: 50–50, right? It's heads. Now I flip it again. What's the probability this time? Of course, it is still 50–50 – the previous outcomes don't in any way influence the next ones and we are confident that the probability of any individual coin flip is always 50–50 regardless of what it was last time. But if I keep doing this and keep getting heads, how many will it take before you are convinced that the coin is biased? There isn't a single 'right' answer to that question, because it depends on what you think of me and whether you've examined the coin. If you trust me and the coin came from your own pocket, then probably quite a lot. If I provided the coin without letting you look at it and you think I'm a bit of a shady character anyway, especially if I asked you to bet and am profiting from the game, then you may quite quickly start to be suspicious that this part of Model Land isn't the place you want to be.

If models are to be used to influence policy decisions, modellers must understand that *trust itself* has a mathematical place in their equations. It isn't a nice-to-have or an afterthought to be pursued by the 'public communication' brigade, it is actually part of the calculation. And it exists in the real world, not in Model Land. That's why the kinds of considerations I have explored throughout this book are so important.

Most people have never been to Model Land

Given the importance of models to the scientific community and the ways in which models now influence so much of our daily lives, it is interesting to note that the general public also remain sceptical about Model Land. Two famous results of behavioural economics are presented as 'paradoxes' that show people choosing to behave irrationally, against the course of action suggested by a model of the situation. Yet I think their results can be explained much more easily by the trivial observation that real people live in the real world, not in Model Land.

First, the Marshmallow Test. This is an experiment where a child is offered a single marshmallow (perhaps left invitingly in front of them on the table), but told that if they can wait they will be given more marshmallows later. Of course, the 'rational' approach, in Model Land, to maximising the number of marshmallows means waiting and taking the larger amount. Predictably, many children take the marshmallow in front of them rather than waiting for more, and this is interpreted variously as lack of willpower or a 'steep discounting rate' of future happiness relative to current happiness. I think this is completely ridiculous. Economists and mathematicians, who live in Model Land professionally, would no doubt wait, but a child does not necessarily expect the stated conditions to be reliable.

In the real world, perhaps their mother will turn up and declare it to be time to leave, or the experiment will turn out to be a trick, or the experimenter will never return, or someone else will come in and say the marshmallow is bad for their teeth and they should have some broccoli instead. If those are possibilities that you attach any significant likelihood to, then of course it is 'rational' to take the marshmallow in front of you, right now. Only some children are likely to accept Model Land and behave accordingly.

Second, the so-called Ellsberg Paradox. I have two containers, which for historical reasons mathematicians always refer to as urns. These urns each contain one hundred red or black balls. One urn has exactly fifty red balls and fifty black ones. The other urn has an unknown number of each colour. You get to pick out one ball (without looking) from one urn, and if it is red, I will pay you £10. Which urn do you wish to choose from? Most people choose the urn that is known to have exactly fifty balls of each colour rather than the one with an unknown number of each colour. This 'irrational' preference is interpreted by behavioural economists as 'uncertainty aversion', since there is still mathematically exactly the same chance of drawing a red ball from either urn and so in Model Land there should be no preference between them. Again, I contend that in the real world there is no equivalence. One urn is known to contain fifty red balls. The other urn could contain anything! How do I know the other urn contains any red balls at all? What if there are some green and blue ones in there? Has the experimenter told me the truth, but not the whole truth? Am I being tricked? In the absence of further evidence, I will certainly choose the first urn.

Very occasionally, we can see the rejection of Model Land directly. A rather frivolous YouGov poll in July 2019 surveyed 2,061 British people, asking whether they would like to take a free trip to the Moon if their safety were guaranteed. Among the half who declined this generous offer, the top reasons were:

- not interested (23%);
- not enough to see/do (11%);
- rather visit other places on Earth (10%);

- 'no point' (9%);
- reject guaranteed safety premise (9%).

The last group, 9% of respondents, directly rejected the conditions attached to the question; in short, they refused to enter Model Land at all and returned to their real-world assumptions about the plausibility of the offer.

Psychologist Mandeep Dhami studies the communication of probability estimates in intelligence analysis. The communication of probability estimates in intelligence communities has been redesigned, following what Dhami describes as a 'major intelligence failure': the misunderstanding of analysts' judgements about the likelihood of existence of weapons of mass destruction in Iraq. The Chilcot Inquiry report, published in 2016, noted that intelligence organisations had made uncertain judgements about the likelihood that Iraq possessed these weapons: that uncertainty was not effectively communicated either to politicians or to the general public. An earlier example was the use of the words 'serious possibility' to communicate the probability of a Soviet invasion of Yugoslavia in 1951. After the Iraq invasion in 2003, new lexicons were developed that identified in numerical terms the probabilities to be associated with phrases like 'very unlikely' and 'virtually certain'. As you might expect, there is disagreement about exactly what numerical ranges should correspond to what phrase. This is a broad and interesting topic in itself, but I want to focus on a single result here. Once the lexicons are chosen and defined (for example: 'unlikely': 15–20%; 'highly likely': 75–85%; and so on), Dhami takes the interesting approach of performing a reverse experiment and asking analysts to identify the numbers that correspond to the lexicon entry itself. Now, you might think this a waste of time – if the lexicon entry says 'unlikely (15–20%)', surely everyone will respond with a probability range of 15–20%? Wrong! In fact, the mean answers given by analysts for the minimum and maximum ends of the 'unlikely (15–20%)' range were 13% and 50%. Similarly, 'highly likely (75–85%)' was identified as 59–92% in practice. A mathematician faced with the same question would have given the trivially obvious answers, but this is not what the real intelligence analysts are doing. It's an incredibly striking and seemingly ridiculous result – even when the number is literally written down in the question, people give a different answer! I interpret this observation to show that the analysts are operating not in Model Land, where the offered probabilities are true, but in the real world, where someone else's estimate is always to be taken with a pinch of salt and most likely overconfident.

That apparent contradiction can also be seen in studies of gambling behaviour, something that you might think is very close to a mathematically ideal situation. While much of mathematical probability theory has been derived from the consideration of idealised gambling games involving dice, cards and coins, it's interesting that gamblers tend not to live in Model Land at all. Partly that's because the casinos live in Model Land, and in Model Land they set the probabilities so that in the long run they will always win, so if you want to get an edge on the house, you have to make use of other kinds of information. Stories are filled with gamblers marking cards, sorting cards, counting spins of the roulette wheel and so on. There is a method for deciding how much to bet when you think you have a better idea of the chance of an outcome than the bookie offering the bet: it's called the Kelly Criterion. In principle, betting according to the Kelly Criterion will maximise your winnings. But in practice, many gamblers escape from Model Land by betting, say, half the Kelly amount, a so-called 'fractional Kelly' strategy which does not assume that you have the right probabilities.

These 'fractional Kelly' gamblers and the intelligence analysts studied by Mandeep Dhami are using the same route of escape from Model Land: simply *downgrade your confidence* arbitrarily but significantly from the Model Land answer. It's a very rough-and-ready exit strategy and not a very graceful one, maybe the equivalent of bashing down a border wall with a large hammer, but it works. This strategy is also used by the IPCC, which has the job of summarising the rapidly growing bank of scientific results about climate change into something semi-digestible by a wider audience. Where it has results available from a set of climate models, it takes an interval in which 90% of modelled results fall and calls it a 'likely (>66%)' range for the real world, another simple but arbitrary confidence downgrade reflecting the fact that the models are not expected to be perfect.

Think outside the box

What I'm emphasising here is that we already have an escape from Model Land that doesn't require you to be a mathematical whizz. Given that real decisions are made in the real world, people have developed real ways of dealing with real uncertainty. We aren't stuck forever, like Buridan's Ass, until we have made a mathematical assessment of exactly which strategy will maximise future happiness. Instead, we just do something. Whether this is motivated by previous experience, different assumptions, advice from others or an emotional conviction about the right thing, we are not paralysed and, though we do not act at random, we do not need to have the best possible answer, only a reasonable one.

Animals and even plants also take actions despite uncertainty, and they don't do it by constructing mathematical models. Instead, a complex interplay of innate and learned qualities help the dog to catch a frisbee without solving any equations of motion or the bird to choose a nesting place without resorting to game theory. They just do what feels right. The computer models we construct are low-dimensional shadows of the multidimensional environment that real organisms exist in. Our proxy indicators for success are pale reflections of the interconnected systems that optimise the long-term reproductive success of individuals, species and ecosystems. From Darwin to Gaia, success is measured in terms of survival and sustainability, and we have yet to see whether the decisions informed by our models will result in long-term survival of the scientific strategy.

Having said all that, you might think that the answer is to throw the models away completely and 'just do what feels right'. I do worry that in the medium term there is a risk of a backlash against models for all of the reasons discussed. You can already see this developing in criticisms of climate models and the policies they have been used to support, and in criticisms of pandemic models and the policies *they* have been used to support. As evolutionary psychologists tell us, though, the things that feel right – like eating sweet foods – may have been great survival strategies for hunter-gatherers but are poor strategies for decision-making in the modern world. Although there are long histories and prehistories of calculative thought, such as the counting of seasons and stars and even the prediction of eclipses, we do not have any evolutionary experience of using computers, spreadsheets and mathematical code to support our thinking. In fact, there tends to be scepticism about grand predictive schemes. Greek stories in particular are full of warnings about what happens when you ask for a prediction of your future, from Oedipus and Cassandra to the Oracle at Delphi, which gave cryptic prophecies that came true in unexpected ways, generally to the dismay of all concerned. Maybe knowing the future is best left to the gods, but complete refusal of model-based information is an unattractive escape from Model Land, especially where the stakes are high. Sometimes we need all the help we can get.

One critical argument in favour of continuing to use mathematical models is their undeniable past successes. Some parts of Model Land are inherently very closely linked to the real world. Models of radioactive decay are accurate to many significant figures. Models of ballistic motion got the Apollo missions to the Moon and back. These kinds of models more or less just represent what we take to be the laws of physics. Models of the weather save billions of dollars every year and many lives. These kinds of models are testable, improvable and ultimately reliable in the sense that we can develop a reasonably good understanding of when they are likely to be right and when they are likely to be wrong.

With that in mind, we should start from a position of confidence that mathematical models and the mathematical language are in themselves useful for representing knowledge about the current state of nature and thinking about the future state of nature. Escaping from Model Land becomes a question of understanding the timescale (or other circumstances) on which our forecasts become less useful and eventually irrelevant. For weather forecasts, we escape from Model Land by noting that tomorrow's forecast generally has quite a high level of skill and is thus sufficient to inform a quite large set of decisions, but that the weather forecast for three weeks ahead has low skill and is thus only sufficient to inform a very limited set of decisions.

Adequate for which purposes?

We've already met the concept of adequacy-for-purpose, as emphasised by Wendy Parker. Parker argues that there is no absolute measure of whether a model is any good or not; we can only assess them with reference to the intended purpose. So weather forecasts might be adequate for the purpose of deciding whether to take a snow shovel in your car tomorrow, but not adequate for the purpose of deciding whether to hold your garden party on a Saturday or Sunday next month.

This framing leaves open the possibility that no model may be adequate for a given purpose. If we wish to decide whether to hold the Winter Olympics in five years' time in one venue or another, we cannot guarantee that either venue will have snow. Fortunately, we have recourse to artificial snow to ensure that the decision and planning can go ahead regardless of the uncertain weather. That's a more elegant way of getting out of Model Land – if your actions are resilient to any of the possible outcomes, you have no need of prediction at all.

On the other hand, it also means that models can develop over time to be adequate for purposes for which they previously were not. Richardson's weather forecast a hundred years ago was rudimentary though impressive. Weather forecasts now – using the same kind of model, but with more data, bigger computers and better algorithms – are incredibly valuable. When computing power is increasing rapidly, there's often an argument for waiting for a better model. It has been joked that a graduate student might have got better results by waiting three years and buying a bigger computer than by spending three years coding and refining their algorithm. The graduate student could probably think of plenty of ways to stay occupied for three years. But although computing power is a practical constraint on what we can do, can we truly say that it is the main limitation on quality? Now that the supercomputers of the mid-1990s can be equalled by pocket devices, are we all doing research that is as important and useful as that performed on those supercomputers? Only in areas where the real world is easily accessible from Model Land – such as weather prediction – can we genuinely make best use of new computing advances. In other fields, model adequacy-for-purpose is as much and generally more limited by our lack of imagination than by raw computational capacity.

So this also relates to debates about whether decisions should be delayed as we search for new information. Should we wait for better predictions of sea-level rise before constructing new barriers? Should we gather data about the virulence of a new disease before reducing travel to an affected area? Should we leave nuclear waste in surface storage while we continue to model the properties of underground caverns? Just as there is no perfect model, there is no perfect decision. In deciding whether to wait for further scientific information to become available, we should also consider how the other information that we have is evolving. If there were more information, it is easy to think that the problem would be solved, but in many cases an improvement of the science or of the models does not actually change the overall decision picture, which includes political disagreement, opportunity costs, and an overall conviction narrative about the situation and how it may evolve in the future. In previous chapters I have argued that the adequacy of a model for its intended purpose is as much to do with how well it functions as an aid to thinking as with its raw predictive capability. If the models themselves are limited by imagination rather than by computing power, we may not have any particular reason to believe that they will necessarily 'improve' in future.

Conversely, though, this offers a potential opportunity to make great improvements to model-informed decision-making despite the constraints of computing performance. If mathematical models can be reimagined as vehicles for creative expression of a range of plausible futures, supporting rather than suppressing social debate about their ethical content, I think they could find a new wave of general interest. From general discussion of climate models to the range of feature articles and social-media threads about how different models have performed during the Covid-19 pandemic, it is clear that there is a huge public interest in understanding more about complex models.

Robust decision-making

Sometimes, staying out of Model Land might mean actively choosing the decisions that are less sensitive to future outcomes. Rather than waiting for better predictions of sea-level rise in order to build flood defences at exactly the right height, for instance, we have a few options:

- build very high barriers now so that they protect well beyond the current range of expected future possibilities;
- build modular barriers that are relatively low-investment now but can be easily added to later;
- enforce fast coastal retreat by compulsory purchase of at-risk property;
- plan slow coastal retreat by stopping development of infrastructure in at-risk zones and replacing it elsewhere; and/or
- make information public and leave a solution to market forces such as unavailability of flood insurance.

The common theme in these cases is taking action to reduce risk which is not optimised to a perfect model prediction of the future (although most strategies do, as you would expect, benefit from having better information). There is a literature on robust decision-making that considers these kinds of trade-offs and alternative strategies in the absence of perfect model information. As RAND Corporation scientist and IPCC author Rob Lempert argues:

> Rather than relying on improved point forecasts or probabilistic predictions, robust decision making embraces many plausible futures, then helps analysts and decision makers identify near-term actions that are robust across a very wide range of futures – that is, actions that promise to do a reasonable job of achieving the decision makers' goals compared to the alternative options, no matter what future comes to pass. Rather than asking what the future will bring, this methodology focuses on what we can do today to better shape the future to our liking.

Lempert and colleagues also share some of my views about the limitations of models and quantitative approaches to decision support:

> Humans also possess various sources of knowledge – tacit, qualitative, experiential, and pragmatic – that are not easily represented in traditional quantitative formalisms. Working without computers, humans can often successfully reason their way through problems of deep uncertainty, provided that their intuition about the system in question works tolerably well.

This use of human decision-making strengths is, as I have shown, key to getting out of Model Land.

Artificial intelligences are not good at robust decision-making: without a human in the loop, they cannot exercise flexibility to account for unanticipated events or assign new value judgements to reflect a changing situation. For this reason, I think it's important that we don't rush to automate all decisions that were previously made by humans, or ossify certain models into inflexible decision-making procedures. The last century of science has moved in the direction of quantification and automation, and in many cases that has been a benefit. But instead of continuing to swing towards a dystopia of full quantification and automation, I think we need to find a balance.

Quantified procedures have been beneficial in decision-making because they make the process more 'objective' and get away from some of the biases of humans. If we leave one person in charge of hiring a new employee, for instance, there are many documented biases that probably result in hiring the person who looks most like the interviewer, rather than the best person for the job. Inclusion of some objective scoring procedures can be beneficial here because it facilitates a more transparent comparison of the qualifications or experience of different candidates, breaking down some barriers. But we can go too far: if we allow a model to select the 'best candidate' by learning from an archive of data about other hiring decisions, it will simply learn to replicate human biases. Similarly, the evolution of business and policy-making decisions away from the so-called 'GOBSAT' method of group decisions by 'Good Old Boys Sat Around the Table' (with all of the non-inclusiveness that the phrase implies, and more) and towards more quantitative methods has been a positive development, but it is also starting to go too far by removing anyone at all from the table and substituting them with what I might call Good Old Boys Sat Behind the Computer. We have not solved the underlying problem here, which has very little to do with the mathematics. Likewise, the replacement of other opaque and unaccountable decision-making structures such as those in criminal sentencing with more transparent criteria has been a positive move, but continuation of this trend all the way to full automation would be extremely concerning.

For this reason, I think that a primary challenge for twenty-first-century decision-making is learning to curb overenthusiasm for mathematical solutions. Much as I love the elegance and simplicity of mathematical descriptions of the world, they are inevitably incomplete and, as I hope I have shown, that incompleteness itself has ethical as well as practical consequences.

Strong objectivity

Objectivity is supposed to be an ideal of scientific practice: science should be objective, the naive argument goes, because we are seeking external truths about nature that are independent of the observer. Our scientific practice therefore should not be contaminated by any kind of bias according to who we are or what we want the result to be. This kind of argument is what underpins replication studies, anonymous peer review, double-blind randomised controlled trials and the importance attached to scientists having no personal investment in the subjects of their study. In this framework, science is a 'view from nowhere' rather than the view of an individual. It is the discovery of empirical fact that can be determined with reference to the external reality, and the scientist is only an impartial medium for that discovery.

But if you have read this far without throwing away the book in disgust, I hope you agree with my view that mathematical models inherently cannot reach this ideal. They represent personal judgements about relative importance. Even where we are making a model that is strictly descriptive and not intended to inform decision-making, it will still be very much dependent on the education, interests, priorities and capabilities of the model-maker. Even if the model is 'made' by an artificial intelligence rather than by a human expert, it inherits in a complex way the priorities of those original creators.

You could take that to mean that models are not scientific, but if you go to that extreme, you will be left with a very thin idea of what does count as science. I think models are scientific, but science is not objective in the naive sense outlined above. Although every individual is stuck with their own subjectivity, I think we can do better by following what philosophers refer to as social accounts of objectivity in science. You may know the parable of blind men meeting an elephant for the first time, which originates in ancient Hindu and Buddhist texts. One man, feeling the tusks of the elephant, declares it to be hard and smooth; another, feeling the tail, declares it to be like a rope; another, feeling the skin, declares it to be rough and warm; another, feeling an ear, declares it to be flappy and thin; another, feeling the trunk, declares it to be strong and thick. All of these are partial descriptions of the elephant, each arrived at from a single, limited perspective. By combining the descriptions and understanding the relative perspective of each contributor, a more comprehensive picture of the whole can be arrived at. Philosopher Sandra Harding has described a system of 'strong objectivity' which maintains that, by approaching a scientific question from different standpoints, we can generate a more effective social kind of objectivity that emerges out of the collaborative work of a diverse group of participants.

For mathematical models, that means embracing alternative models produced by different experts as equally valid descriptors, although they may disagree about representation and attach completely different levels of importance to the different elements. We do not want to overfit an elephant using too many parameters when we have a limited perspective on its characteristics. This is how we can mitigate what Adichie identified as 'the danger of a single story'. You will not *like* all of the stories; it is a matter of personal taste. Some you may think to be facile or boring. Other people may find your own favourites stilted or incomprehensible. Just as the expanding canon of literature provides entry points for many different people to comprehend internal lives and perspectives beyond their own, so an expanding canon of models can do the same for mathematical understanding of physical and social worlds. We do not need to cancel Shakespeare or delete our numerical climate models to understand the immense value of introducing other perspectives to a wider audience.

On getting the right answer(s)

What I hope I have shown in this book is that Model Land is not an objective mathematical reality, but a social idea. There are many Model Lands, and each one takes on the social context, assumptions, expectations and prejudices of its creator. Models shape the way that we frame our scientific questions, like the economic models that recast all decisions into financial terms. Models shape the way that we think about possible futures, like the climate change models that visualise certain kinds of outcomes. And models shape the kinds of actions that we can conceive of taking to influence those futures, like the epidemiological models that allow a limited set of possible interventions. Mathematical models are fundamental to high-impact decisions in spheres that affect us all. My examples have been largely drawn from economic, public health and environmental decision-making, but these represent only a small fraction of the domains in which modelling has become more and more influential.

A successful model may be one that helps us to make better decisions, but we are first going to have to define what we mean by 'better decisions'. A better decision, presumably, is one that is more in line with our declared values. For example, if we care solely about global GDP, then a model that allows us to take real-world decisions to attempt to maximise global GDP is a successful model. If we care about two things that are somewhat in conflict, then a successful model is one that helps us to describe that trade-off in a way that will allow decision-makers to balance the two. But successful models are not just good predictors, though a good predictor can be a useful model. As we have seen, models are also metaphors that facilitate communication, frame narratives and include value judgements with scientific information.

Although the whole of society should have some responsibility for value judgements, it is not plausible for the whole of society to be directly involved in model-building or in model criticism. For that, we need experts who have both the technical fluency to create models and the domain knowledge of the system being modelled. I have suggested that science would be both more effective and more trustworthy, and likely also more trusted, if those experts were selected to maximise the diversity of perspectives rather than to optimise for a certain limited kind of expertise. To do that would require a sea change in the way that our knowledge systems are structured, and I am hardly optimistic about the wider prospects

– but every individual modelling choice can be made with the importance of this diversity in mind, if those who fund and guide model development desire it.

Mathematical modelling is a powerful tool for constructing narratives about future actions and the costs and benefits of different outcomes. As such, it well deserves its place at the top table of political decision-making. I want to emphasise that this is not a slippery slope to some postmodern theory of model relativism, in which anything goes and all models are equally valid. But I hope that my illustration of the subjectivity of modelled knowledge, and its dependence on the personal and political values and perspectives of the humans involved in creating it, signposts some necessary improvements. There is no single right answer, and so there are opportunities for abuse and manipulation, especially where the veneer of science is used to hide value judgements. We need an ethical framework for mathematical modelling.

The point is not to throw away the insight that has been gained by modelling so far, but to reinforce the foundations of the model as an edifice rather than to continue to balance more and more blocks on top of its shaky spires. The only plausible foundations are either detailed discussion and agreement on the value judgements that underlie complex models, or clear explanation of the different value judgements that result in different answers. Otherwise we will continue to be lost in a world where differing value judgements are dressed up as scientific disagreements, like the disagreements about burning fossil fuels in order to promote economic growth, or closing down sectors that support livelihoods in order to save lives from a virus. It is spectacularly unproductive to direct the emotional and moral indignation that these kinds of questions quite reasonably attract into some sort of 'scientific' debate. There simply is no right answer. The question of how we should act in the world is always a moral question. Doing the mathematics in a defensible way is an important stepping stone, although there are other bridges elsewhere. We can only come to some sort of social and political consensus by acknowledging the contingency of our models on value judgements, discussing those value judgements and then navigating the difficult waters of compromise.

Drawing on the ideas and examples I have already presented, here are five principles for responsible modelling, which I hope can help form a code of practice for the working modeller or assist the non-modeller in asking searching questions of those who do research on their behalf.

1. Define the purpose

I have referred a few times to the 'adequacy-for-purpose' perspective of Wendy Parker. As a starting point for creating models, we need to decide to what purpose(s) they are supposed to be put. If we want to inform decisions about a certain group of people, they should be represented. If we want to consider certain courses of action, they should be modellable. If the decisions will involve trading off different goods or benefits, they should all in principle be considered by the model.

In practice, that's very difficult to achieve. Most models are not adequate for the purpose of making any decision, although they may be adequate for the purpose of informing the decision-maker about some part of the decision, along with other models that consider other parts. So, for instance, an epidemiological model can inform us about viral transmission and hospital pressure, but not about the economic effects of closing businesses. Where the model is only going to be a small part of the information going into the decision, that needs to be clear. The boundaries and remit of the model should not be oversold.

Ask the modeller: What kinds of questions can this model answer? What kinds of questions can it not answer? What other information would need to be added?

2. Don't say 'I don't know'

If all models are wrong, the future is unknowable and subject to radical uncertainty, and the models are at the mercy of subjective and potentially biased 'expertise', we might ask whether there is any point in making models about uncertain futures? Do we need to go to Model Land at all? If you have read John Kay and Mervyn King's book, with its emphasis on being able to say 'I don't know' when confronted with radical uncertainty, perhaps you feel downcast by the failure of predictive science to rise to the challenges of the twenty-first century. If you take a purely mathematical view of the challenges involved in modelling, you may come to the conclusion that 'I don't know' is the right answer.

But I hope that my more social account of the uses and limitations of models has shown that there are insights to be gathered from even the most limited and partial of models. If we can give up on the prospect of perfect knowledge and let go of the hope for probabilistic predictions, we are not back to square one – there are what Kay and King would call alternative narratives in each model, which in themselves contain useful insights about the situation, although each one is partial and biased. We know nothing for certain, but we do not know nothing. Don't abdicate responsibility by saying 'I don't know'.

Ask the modeller: If the model can't predict quantitatively, what else can it show us about the situation? Do these qualitative insights agree or disagree with other models and observations? What alternative stories can it tell about the future? Can we make decisions that are less sensitive to the different plausible outcomes?

3. Make value judgements

All models require value judgements. If you cannot find any in your model, look harder or ask someone else to look for you, preferably someone with a non-technical background who is directly affected by the decision informed by your model. They will find some.

When you understand the value judgements you have made, write about them. Why are they made in this way? Is there anyone who might disagree? What alternative specifications would be possible if you made different value judgements, and are the outcomes of the model sensitive or robust to the alternative specifications? If it isn't clear to everyone that we 'have to' flatten the pandemic curve or 'have to' reduce greenhouse gas emissions, that's probably not because they are stupid but because they disagree with the value judgements inherent in the question and the model. Allow for representations of alternative judgements without demonising those that are different from your own.

In a very boringly ideal world, there might be consensus on values. If, as is more likely, there are differences of opinion, we need to prioritise the exploration of differences in value systems just as much as the exploration of tweaking different technical parameters.

Ask the modeller: What makes this a 'good' model? What constitutes a 'good' outcome for this model? If decisions informed by this model will influence other people or communities, were they consulted or engaged in the making of the model? Is there consensus on the underlying value judgements or are they a matter of political debate?

4. Write about the real world

When you're explaining your results to somebody else, get out of Model Land and own the results. You're the expert: in what ways is this model inadequate or misinformative? What important processes does it fail to capture? Do you, personally, believe that the projections of the model are likely to be reasonably accurate, or do you take it only as a guide to qualitative behaviours?

When writing a scientific paper about your model, you might like to report the results verbatim as a report from Model Land. But if you are communicating with non-experts who do not live in (your) Model Land, you will need to translate. Do you have high or low confidence in the numbers you are reporting? What assumptions or preconditions would need to be true for your statements to hold?

Ask the modeller: Is this a Model Land result or a real-world result? Given that the model is not perfect, whose judgement has been used to translate from Model Land into the real world? Might someone else make different judgements about quality? Can the assumptions and preconditions be tested directly?

5. Use many models

Following Sandra Harding's framework of strong objectivity, gathering insights from as diverse a range of perspectives as possible will help us to be maximally informed about the prospects and possibilities of the future. In particular, it is easy to identify where someone unlike you might have political views or personal circumstances that influence their judgement. Embrace these alternatives – especially if it is emotionally difficult – and consider what they might identify in you. How do the unspoken assumptions of your personal or educational background influence the priorities of your model? Are you pretending to be completely disinterested?

Ask the modeller: What other models of this situation exist? How might someone with a different disciplinary background or different personal or political interests create a different model? Are you encouraging diversity or suppressing it?

'Here Be Dragons': at the borders of Model Land

Although all models are wrong, many are useful. I have given a different account of generating insights from models, one that builds on the human and subjective nature of models as metaphors and narratives, rather than 'scientific' windows onto some external truth. In order to transfer insights from Model Land back to the real world in which we live, there are two options that must be combined in different measures according to the type of problem under consideration. One way to escape from Model Land is through the quantitative exit, but this can only be applied in a very limited set of circumstances. Almost all of the time, our escape from Model Land must use the qualitative exit: expert judgement. This necessary reliance on expert judgement presents us with challenges: who is an expert? Why should we trust them? Are they making assumptions that we can agree with? And if models are, as I have argued, very much engines of scientific knowledge and social decision-making rather than simple prediction tools, we also have to consider how they interact with politics and the ways in which we delegate some people to make decisions on behalf of others. The future is unknowable, but it is not ungraspable, and the models that we create to manage the uncertainty of the future can play a big role in helping to construct that future. As such, we should take the creative element of metaphor and narrative in mathematical models at least as seriously as their predictive accuracy.

Model Land provides us with maps of the future, but they are not always the maps that we need. Some are plain and simple; some are elaborate and embellished; most are in some ways misleading; all were drawn by someone fallible; all have limits beyond which we can only write 'Here Be Dragons'. The terrain outside Model Land is bumpy and disputed. While our models can predict, warn, motivate or inspire, we must ourselves navigate the real-world territory and live up to the challenge of making the best of our imperfect knowledge to create a future worth living in.

Acknowledgements

What a time to write a book about mathematical models. Between the time I started to write in 2019 and when I finally completed the text in 2022, a large-scale case study of model use to support actions during the Covid-19 pandemic has played out in real time. I hope this book is therefore a timely intervention which will inform other discussions about mathematical modelling and its role in society.

This book has only been possible thanks to the support of the London Mathematical Laboratory (LML), which enabled me to write part time, so that I did not have to devote evenings or weekends to it. I'm especially grateful for the confidence of Alex Adamou (then-Director of LML), who encouraged me to go ahead and do it, helped put me in touch with publishers and offered much-needed moral support at intervals. Thanks also to Ole Peters for support and for sharing some of the trials and tribulations of writing a book. Kame Boevska and Chloe Scragg deserve a mention for much-appreciated logistical support. I look forward to LML going from strength to strength under the new leadership of Colm Connaughton.

My intellectual debts are varied and I hope many are acknowledged already in the text, but in particular I have to thank Leonard Smith. Lenny has been talking and writing about mathematical models for many years, and while I have taken a slightly different slant on them here, I think all of the foundational mathematical insights are ones I learned from him. I owe to him the concept of Model Land, and also the title of this book, taken from a paper we co-wrote in 2019. Lenny's group at the LSE, the Centre for the Analysis of Time Series (CATS), was a stimulating and exciting place to work from 2012 to 2020 and introduced me to many other interesting uses of mathematical modelling. As a co-lead of the CRUISSE Network, Lenny also brought me into contact with ideas about the social context of modelling – including David Tuckett's conviction narrative theory – which developed my own thinking and are reflected here. Thank you, Lenny, for your generous insights, your focus on mathematical and real-world interests rather than narrow academic kudos, and your friendship.

Other members of CATS have also been supportive colleagues: thanks to Dave Stainforth, Ed Wheatcroft, Hailiang Du, Hannah Nissan, Henry Wynn, Ewelina Sienkiewicz, Emma Suckling, Lyn Grove and especially Jill Beattie, who helped to untangle the whole thing in 2020.

I was very fortunate to be awarded a UKRI Future Leaders Fellowship in 2021, which enabled me to dedicate more time in the final few months to finishing this book rather than writing other proposals. Thank you to the UKRI reviewers and panellists who were so enthusiastic about my proposal, for their recognition of the importance of interdisciplinary work and for not being put off by my jack-of-all-trades lack of expertise in some key areas. Mary Morgan, Roman Frigg and Arthur Petersen very kindly supported that proposal and I'm so grateful for their help and advice here and elsewhere. The project builds on the concepts outlined in this book, so I will no doubt have a completely different perspective in a few years' time and will have to rewrite it :)

Huge thanks also to Marina Baldissera Pacchetti, Julie Jebeile and other members of the Climate Change Adaptation, Vulnerability & Services reading group on alternative epistemologies of climate science. The sharpness of their insights is only matched by their warmth and generosity. This book is much better for our conversations, although I know that it also reflects how much more work I still need to do. I'm grateful for the introduction to so many important topics (and sorry for not using the F-word).

I have benefited from Western white middle-class privilege, from my formal education and background in science. These have also enabled me to write this book and they contribute to making it this book and not a different book. I am also limited by that background and I appreciate the generosity of readers in tolerating the errors, bad takes and misinterpretations that I have no doubt committed to paper here as a result of my ignorance of other perspectives.

Writing has been a rather exhausting learning experience on all fronts. I appreciate the efforts of editors Sarah Caro and T. J. Kelleher in pushing me to finish the job. My thanks to all who read partial or full drafts, including Mark Buchanan, Jag Bhalla, Nadia Farid and Lenny Smith. The moral support of many lovely friends and neighbours has been invaluable. Thank you all, especially to Silvia, Wyc, Peni, Stef, Paul, Sarah, Jacqui and Tom. My parents and wider family have also been supportive, even though during the pandemic we have not seen as much of each other as we would like. My lovely children have made writing this book somewhat harder and longer than it might otherwise have been, but I hope it is also better for the efforts. And my husband, Chris, has made it possible for me to write despite all of the other commitments in our lives, and also talked me out of despairing, quitting or taking it too seriously. Thank you, Chris.

Further Reading

I thought it would be useful to provide further reading grouped by chapter. This is not a comprehensive list of sources, but if you are interested, here's where to follow up my references and find out more.

Chapter 1: Locating Model Land

King, Mervyn, and John Kay, *Radical Uncertainty: Decision-Making for an Unknowable Future*, Bridge Street Press, 2020

Chapter 2: Thinking Inside the Box

#inmice, https://twitter.com/justsaysinmice
https://www.climateprediction.net
Held, Isaac, 'The Gap Between Simulation and Understanding in Climate Modeling', *Bulletin of the American Meteorological Society*, *86*(11), 2005, pp. 1609–14
Mayer, Jurgen, Khaled Khairy and Jonathon Howard, 'Drawing an Elephant with Four Complex Parameters', *American Journal of Physics*, *78*, 2010
Morgan, Mary, *The World in the Model*, Cambridge University Press, 2012
Page, Scott, *The Model Thinker: What You Need to Know to Make Data Work for You*, Basic Books, 2019
Parker, Wendy, 'Model Evaluation: An Adequacy-for-Purpose View', *Philosophy of Science*, *87*(3), 2020
Pilkey, Orrin, and Linda Pilkey-Jarvis, *Useless Arithmetic: Why Environmental Scientists Can't Predict the Future*, Columbia University Press, 2007

Stainforth, David, Myles Allen, Edward Tredger, and Leonard Smith, 'Confidence, Uncertainty and Decision-Support Relevance in Climate Predictions', *Philosophical Transactions of the Royal Society A: Mathematical, Physical and Engineering Sciences*, *365*(1857), 2007

Stoppard, Tom, *Arcadia*, Faber & Faber, 1993

Chapter 3: Models as Metaphors

Adichie, Chimamanda Ngozi, 'The Danger of a Single Story', TED talk (video and transcript), 2009

Bender, Emily, Timnit Gebru, Angelina McMillan-Major and Shmargaret Shmitchell, 'On the Dangers of Stochastic Parrots: Can Language Models Be Too Big?', *Proceedings of the 2021 ACM Conference on Fairness, Accountability, and Transparency*, 2021

Bolukbasi, Tolga, Kai-Wei Chang, James Y. Zou, et al., 'Man Is to Computer Programmer as Woman Is to Homemaker? Debiasing Word Embeddings', *Advances in Neural Information Processing Systems*, *29*, 2016

Boumans, Marcel, 'Built-in Justifications', in Morgan and Morrison (eds), *Models as Mediators*, 1999, pp. 66–96

Daston, Lorraine, and Peter Galison, *Objectivity*, Princeton University Press, 2021

Frigg, Roman, and Matthew Hunter (eds), *Beyond Mimesis and Nominalism: Representation in Art and Science*, Springer, 2010

Morgan, Mary, and Margaret Morrison, *Models as Mediators*, Cambridge University Press (Ideas in Context Series), 1999

O'Neil, Cathy, *Weapons of Math Destruction: How Big Data Increases Inequality and Threatens Democracy*, Penguin, 2016

Svetlova, Ekaterina, *Financial Models and Society: Villains or Scapegoats?*, Edward Elgar, 2018

Chapter 4: The Cat that Looks Most Like a Dog

Berger, James, and Leonard Smith, 'On the Statistical Formalism of Uncertainty Quantification', *Annual Review of Statistics and its Application*, *6*, 2019, pp. 433–60

Cartwright, Nancy, *How the Laws of Physics Lie*, Oxford University Press, 1983

Hájek, Alan, 'The Reference Class Problem is Your Problem Too', *Synthese*, *156*(3), 2007

Mayo, Deborah, *Statistical Inference as Severe Testing: How to Get Beyond the Statistics Wars*, Cambridge University Press, 2018

Taleb, Nassim, *Fooled by Randomness: The Hidden Role of Chance in Life and in the Markets*, Penguin, 2007

Thompson, Erica, and Leonard Smith, 'Escape from Model-Land', *Economics*, *13*(1), 2019

Chapter 5: Fiction, Prediction and Conviction

Azzolini, Monica, *The Duke and the Stars*, Harvard University Press, 2013

Basbøll, Thomas, *Any Old Map Won't Do; Improving the Credibility of Storytelling in Sensemaking Scholarship*, Copenhagen Business School, 2012

Davies, David, 'Learning Through Fictional Narratives in Art and Science', in Frigg and Hunter (eds), *Beyond Mimesis and Convention*, 2010, pp. 52–69

Gelman, Andrew, and Thomas Basbøll, 'When Do Stories Work? Evidence and Illustration in the Social Sciences', *Sociological Methods and Research*, *43*(4), 2014

Silver, David, Thomas Hubert, Julian Schrittwieser, et al., 'A General Reinforcement Learning Algorithm That Masters Chess, Shogi, and Go Through Self-Play', *Science*, *362*(6419), 2018

Tuckett, David, and Milena Nikolic, 'The Role of Conviction and Narrative in Decision-Making under Radical Uncertainty', *Theory and Psychology*, 27, 2017

Chapter 6: The Accountability Gap

Birhane, Abeba, 'The Impossibility of Automating Ambiguity', *Artificial Life*, 27(1), 2021, pp. 44–61

——, Pratyusha Kalluri, Dallas Card, William Agnew, et al., 'The Values Encoded in Machine Learning Research', arXiv preprint arXiv:2106.15590, 2021

Pfleiderer, Paul, 'Chameleons: The Misuse of Theoretical Models in Finance and Economics', *Economica*, 87(345), 2020

Chapter 7: Masters of the Universe

Alves, Christina, and Ingrid Kvangraven, 'Changing the Narrative: Economics after Covid-19', *Review of Agrarian Studies*, 2020
Ambler, Lucy, Joe Earle, and Nicola Scott, *Reclaiming Economics for Future Generations*, Manchester University Press, 2022
Ayache, Elie, *The Blank Swan – The End of Probability*, Wiley, 2010.
Derman, Emanuel, *Models. Behaving. Badly.: Why Confusing Illusion with Reality Can Lead to Disaster, on Wall Street and in Life*, Free Press, 2012
Frydman, Roman, and Michael Goldberg, *Beyond Mechanical Markets*, Princeton University Press, 2011
Haldane, Andrew, 'The Dog and the Frisbee', speech at Jackson Hole, Wyoming, 31 August 2012
Lowenstein, Roger, *When Genius Failed: The Rise and Fall of Long Term Capital Management*, Fourth Estate, 2002
MacKenzie, Donald, *An Engine, Not a Camera: How Financial Models Shape Markets*, MIT Press (Inside Technology Series), 2008
March, James, Lee Sproull and Michal Tamuz, 'Learning from Samples of One or Fewer', *Organization Science*, 2(1), 1991
Rebonato, Riccardo, *Volatility and Correlation*, John Wiley, 1999
Stiglitz, Joseph, 'Where Modern Macroeconomics Went Wrong', *Oxford Review of Economic Policy*, 34, 2018
Taleb, Nassim, *The Black Swan: The Impact of the Highly Improbable*, Random House, 2007
Wilmott, Paul, and David Orrell, *The Money Formula: Dodgy Finance, Pseudo Science, and How Mathematicians Took Over the Markets*, Wiley, 2017
——, and Emanuel Derman, 'The Financial Modelers' Manifesto', https://wilmott.com/financialmodelers-manifesto/, 2009

Chapter 8: The Atmosphere is Complicated

Allen, Myles, Mustafa Babiker, Yang Chen, et al., 'IPCC SR15: Summary for Policymakers', in *IPCC Special Report: Global Warming of 1.5C*, Intergovernmental Panel on Climate Change, 2018

Anderson, Kevin, 'Duality in Climate Science', *Nature Geoscience*, *8*(12), 2015

Beck, Silke, and Martin Mahony, 'The Politics of Anticipation: The IPCC and the Negative Emissions Technologies Experience', *Global Sustainability*, *1*, 2018

Burke, Marshall, Solomon Hsiang and Edward Miguel, 'Global Non-Linear Effect of Temperature on Economic Production', *Nature*, 527(7577), 2015

Edwards, Paul, *A Vast Machine: Computer Models, Climate Data, and the Politics of Global Warming*, MIT Press, 2010

Hänsel, Martin C., Moritz A. Drupp, Daniel J. A. Johansson, et al., 'Climate economics support for the UN climate targets', *Nature Climate Change* *10*(8), 2020

Hayles, N. Katherine, *Unthought*, University of Chicago Press, 2020

Hourdin, Frédéric, Thorsten Mauritsen, Andrew Gettelman, et al., 'The Art and Science of Climate Model Tuning', *Bulletin of the American Meteorological Society*, *98*(3), 2017

Hulme, Mike, *Weathered: Cultures of Climate*, SAGE Press, 2016

IPCC, *Climate Change 2022: Mitigation of Climate Change. Contribution of Working Group III to the Sixth Assessment Report of the Intergovernmental Panel on Climate Change*, ed. P. R. Shukla, J. Skea, R. Slade, et al., Cambridge University Press, 2022, doi: 10.1017/9781009157926

Lloyd, Elisabeth, and Eric Winsberg (eds), *Climate Modelling: Philosophical and Conceptual Issues*, Palgrave Macmillan, 2018

Lorenz, Edward, 'Predictability: Does the Flap of a Butterfly's Wings in Brazil Set Off a Tornado in Texas?', *American Association for the Advancement of Sciences*, 1972

——, *The Essence of Chaos*, University of Washington Press, 1993

Lynch, Peter, *The Emergence of Numerical Weather Prediction: Richardson's Dream*, Cambridge University Press, 2006

McCollum, David, Ajay Gambhir, Joeri Rogelj and Charlie Wilson, 'Energy Modellers Should Explore Extremes More Systematically in Scenarios', *Nature Energy*, *5*(2), 2020

McLaren, Duncan, and Nils Markusson, 'The Co-evolution of Technological Promises, Modelling, Policies and Climate Change Targets', *Nature Climate Change*, *10*(5), 2020

Palmer, Tim, and Bjorn Stevens, 'The Scientific Challenge of Understanding and Estimating Climate Change', *Proceedings of the National Academy of Sciences*, *116*(49), 2019

Petersen, Arthur, *Simulating Nature: A Philosophical Study of Computer-Simulation Uncertainties and their Role in Climate Science and Policy Advice*, Het Spinhuis, 2006

Pulkkinen, Karoliina, Sabine Undorf, Frida Bender, et al., 'The Value of Values in Climate Science', *Nature Climate Change*, 2022

Quiggin, Daniel, Kris De Meyer, Lucy Hubble-Rose and Antony Froggatt, *Climate Change Risk Assessment 2021*, Chatham House Environment and Society Programme, 2021

Smith, Leonard, *Chaos: A Very Short Introduction*, Oxford University Press (Very Short Introduction Series), 2007

Winsberg, Eric, *Philosophy and Climate Science*, Cambridge University Press, 2020

Chapter 9: Totally Under Control

D'Ignazio, Catherine, and Lauren Klein, *Data Feminism*, MIT Press, 2020

Erikson, Susan, 'Global Health Futures? Reckoning with a Pandemic Bond', *Medicine Anthropology Theory*, *6*(3), 2019

Ferguson, Neil, Derek Cummings, Christophe Fraser, et al., 'Strategies for Mitigating an Influenza Pandemic', *Nature*, *442*(7101), 2016

Hine, Deirdre, *The 2009 Influenza Pandemic: An Independent Review of the UK Response to the 2009 Influenza Pandemic*, UK Cabinet Office, 2010

Howlett, Peter, and Mary Morgan (eds), *How Well Do Facts Travel? The Dissemination of Reliable Knowledge*, Cambridge University Press, 2010

Mansnerus, Erika, *Modelling in Public Health Research: How Mathematical Techniques Keep us Healthy*, Springer, 2014

Neustadt, Richard, and Harvey Fineberg, *The Swine Flu Affair: Decision-making on a Slippery Disease*, US Dept of Health, Education and Welfare, 1978

Rhodes, Tim, Kari Lancaster and Marsha Rosengarten, 'A Model Society: Maths, Models and Expertise in Viral Outbreaks', *Critical Public Health*, *30*(3), 2020

——, and Kari Lancaster, 'Mathematical Models as Public Troubles in COVID-19 Infection Control: Following the Numbers', *Health Sociology Review*, 29(2), 2020

Richardson, Eugene, 'Pandemicity, COVID-19 and the Limits of Public Health Science', *BMJ Global Health*, 5(4), 2020

Spinney, Laura, *Pale Rider: The Spanish Flu of 1918 and How it Changed the World*, Vintage, 2018

Chapter 10: Escaping from Model Land

Dhami, Mandeep, 'Towards an Evidence-Based Approach to Communicating Uncertainty in Intelligence Analysis', *Intelligence and National Security*, 33(2), 2018

Harding, Sandra, *Objectivity and Diversity: Another Logic of Scientific Research*, University of Chicago Press, 2015

Marchau, Vincent, Warren Walker, Pieter Bloemen and Steven Popper (eds), *Decision Making under Deep Uncertainty*, Springer, 2019

Scoones, Ian, and Andy Stirling, *The Politics of Uncertainty: Challenges of Transformation*, Taylor & Francis, 2020

Index

BASIC
BOOKS